BLIND
AMBITION

How to Go from Victim to Visionary

CHAD E. FOSTER

HarperCollins
Leadership

AN IMPRINT OF HARPERCOLLINS

Published by HarperCollins Leadership, an imprint of HarperCollins Focus LLC.

Any internet addresses, phone numbers, or company or product information printed in this book are offered as a resource and are not intended in any way to be or to imply an endorsement by HarperCollins Leadership, nor does HarperCollins Leadership vouch for the existence, content, or services of these sites, phone numbers, companies, or products beyond the life of this book.

ISBN 978-1-4002-2265-0 (eBook)

ISBN 978-1-4002-2264-3 (HC)

Library of Congress Control Number: 2020949802

Printed in the United States of America

20 21 22 23 LSC 10 9 8 7 6 5 4 3 2 1

To Evie, for everything. I mean everything!
The love, the sacrifices, and support.

To Juliana, for your open heart, genuine personality,
and relentless effort.

To Jackson, for your fearless spirit, caring soul,
and unwavering determination.

To Mom, who showed me the meaning of true love—selflessness.

To Dad, who taught me how to be tough enough
and never settle for less.

CONTENTS

BLIND AMBITION: HOW TO GO FROM VICTIM TO VISIONARY

THE BLACK SUV ROLLED THROUGH the hills of Northern Virginia as I sat in the back seat, headed to my big job interview. I'd been fielding calls from Ben, my interviewer, for many months about possibly coming to work for his company, SRA International. In recent years, while working for another big tech company, I'd developed a special niche in business analytics, becoming an expert in the arcane field of financial modeling and pricing strategies. Now it was time for me to make a move, and I felt sure I could be of help to SRA.

The SUV glided to a stop in front of SRA's offices, and I stepped out into the warm August sunshine. I grabbed my black leather briefcase, made my way to the rear of the vehicle, and popped open the liftgate. Romeo, my 115-pound German shepherd guide dog, was waiting patiently for me, furiously wagging his tail. I collected his leash and called him out of the vehicle. Once he was standing at attention by my left side, I gave him the "Door!" command, and he led us to the building entrance.

Romeo was an imposing figure, with a velvety black and tan coat and a head the size of a basketball. He was both adorable and intimidating. Romeo and I were fast walkers, and from afar, you might not think I was Romeo's blind handler. With me in a business suit and dark sunglasses, the two of us appeared more like a plainclothes police officer walking a K-9 service dog.

Once inside SRA's headquarters, Ben introduced me around to other people on his team. Then we returned to his office, where Romeo led me to a chair. Once I sat down, Romeo could relax because his work was done for the moment. He remained perfectly quiet and motionless for the duration of the meeting, like any well-trained guide dog.

In our prior conversations, Ben and I had discovered that we spoke a common language regarding how large-scale technology services should be priced and how the financials of those deals could be best managed. Now, for the first time, I was able to open up my laptop on his desk and show him some of the computer models I'd developed to inform my insights. I could tell by the sound of Ben's voice that he was impressed.

For the next several minutes, Ben explained how the role he had in mind for me required much more than mere technical expertise. I would not only have to analyze vast amounts of data to arrive at my recommended pricing guidance, I'd also have to convince the company's executives that my strategy was the winning and profitable approach. "These are multibillion-dollar deals, you understand," Ben said. "Some of them can make or break the company."

Ben paused. In a tone that reflected his natural curiosity, he asked, "Chad, can you really do all these things?"

I took only a split second to reply. "Ben," I said, "this stuff is so easy, I can do it with my eyes closed."

I still remember the sound of Ben bursting into laughter. He hired me later that month, mainly because he knew I could do the job, but also because he knew I was comfortable enough in my own skin to joke about being blind. I had the business and tech skills to succeed, but just as important, he saw firsthand how I could handle myself in a meeting with self-confidence and good humor.

I've learned to find humor in being blind, because I can't change how blindness is such a big part of my life. Wherever I go, I'm led by a German shepherd guide dog (it is not easy to hide a large German shepherd in a conference room). So, because it's nearly impossible for me to pretend to be someone I'm not, I decided many years ago to accept who I am and deal with my everyday reality head-on—embracing it, owning it, and yes, even welcoming it. It won't help me or anyone else if I'm ashamed or apologetic about being blind, or if I fail to express myself fully because of it. This self-awareness was forced upon me by my blindness, but it has made me a better person today than I was before I lost my eyesight. I am a better person because of my blindness—not in spite of it.

As someone who could see until young adulthood, I know how tempting it is for many of us to put up a false front and avoid the responsibility of sincere self-acceptance and authentic self-expression. My blindness disallows the charades and pretensions I once hid behind. I have learned in the years since losing my eyesight to live according to my own standards and on my terms.

For those of you not as fortunate, this is my story.

ONE

THE FINAL
SUNSET

"THIS CAN'T BE HAPPENING."

The brightly lit pages of the textbook on my desk had just dissolved into a muddled mass of colors. My eyes were no longer working. In the course of several minutes, what looked like a dense swarm of bees had overtaken my field of vision.

"Not this. Not now."

I muttered to myself with frustration. How would I explain why I hadn't completed my assignment?

It was a cool East Tennessee evening in the fall of 1996, just a few months shy of my twenty-first birthday, when I realized I would soon permanently lose my ability to see.

The trees on my college campus that afternoon were awash in fall foliage—deep crimson red, brilliant pumpkin orange, and sawdust shades of yellow-brown. Outside the suburban Knoxville home I'd been raised in, the yard was filled with brightly colored, fallen leaves. But in my second-floor bedroom, my life was taking a much darker turn.

I was hovering over my desk, with a blazing-bright lamp beaming down on my college literature assignment, as I tried to blink away the swarms of "floaters" obscuring the words on the page. Floaters—those dark little specks that most people experience as fleeting shadows in their field of vision—had been an increasing problem for me of late. I was having more and more bad days with them, a common symptom of my eye condition, retinitis pigmentosa. Without warning, the floaters would start piling up before me, like swarming bees. I'd have to stop what I was doing and roll my eyes to make the floaters clear out, just to get a few moments of visual clarity.

But until this moment, the floaters had never prevented me from getting my schoolwork done. What should have been a thirty-minute reading assignment was taking me more than two hours, and I still wasn't finished. Feelings of frustration welled up inside me. No matter how frequently I rolled my eyes and rubbed them, all I could see was a teeming mosh pit of floaters. The textbook in front of me dissolved into a sea of black and white bubbles.

I grew increasingly impatient and tried every trick to work around the problem. Now nothing I tried was of help. No matter how close I looked or how much I tried to brighten the page, I could not make out any of the words. Finally giving into my frustration, I wadded up my notes into a ball and threw it against the wall.

Enraged, I shoved my books off my desk and onto the floor and screamed, *"Damn it!!"*

When my rage was finally exhausted, I sat there trembling, with a cold feeling of sadness that dripped down into the deepest crevices of my soul. Weighed down by the enormous gravity of my new reality, it was hard to move. Hard to breathe.

"This is it," I said to myself. This was what the doctors predicted. The blindness that had been closing in on me slowly since I was a small child had arrived. I was struck with the disturbing realization that very soon I would be reading my last page of print. I'd see my last smiling face. I'd scribble my last note. I'd watch my final football game. I'd enjoy my final sunset. All the gifts that vision grants us would soon be lost to me forever.

I looked around the room and could barely make out the details of my wall decorations. I mostly saw only hazy regions of light and dark, blurred by the tears that were filling my eyes.

Life as I'd known it was coming to an end. I was completely unprepared for what would come next.

MAGICAL THINKING

We were warned this day could arrive nearly twenty years earlier. My parents noticed how their three-year-old toddler kept bumping into things, especially in dimly lit rooms. They took me to eye specialists at Duke University Medical Center, and it didn't take long for them to diagnose me with retinitis pigmentosa. RP, as it's called, is the medical term for a group of rare genetic disorders that cause a gradual and irreversible breakdown of the photoreceptor cells in the retina. Globally, it affects about one in every four thousand people.

Early symptoms of RP typically include difficulty seeing at night and a loss of side vision or peripheral vision. At the age of three, I had both of these symptoms. The doctors told us that I could go blind eventually, possibly at an early age, although some people with RP don't lose their eyesight until middle age or later, and a lucky few retain minimally functioning eyesight their entire lives. We were cautioned that there was no real cure for RP and no effective means of treating it or slowing its progression.

My parents cried the entire drive home. It's hard to imagine what it must have been like to hear the news about their three-year-old child. The doctor hadn't given them much hope to hang onto. "Try to prepare for the day when Chad's vision is gone," he'd said. "In the meantime, enjoy life while you can."

For the next seventeen years, I followed only the second half of the doctor's advice. I enjoyed my life, but I never spent a minute preparing to be blind.

I grew up in a modest single-family home with my parents and my older brother, Erick, in the suburban Knoxville town of Halls. My father, Charles, was a hardworking car salesman who could sell ice in Antarctica. My mother, Peggy, worked as a bookkeeper most of her life and brought that spirit of care and buttoned-up organization to our everyday lives. Everyone on both sides of my family had great eyesight, with no history of RP symptoms. Doctors later discovered that my parents each happened to be carriers of a hidden, or "recessive," gene for RP. That meant there was only a one-in-four chance that their offspring would get RP. I was the unlucky winner of that lottery.

And yet, I behaved like any other adventurous little kid. I was full of energy and determined to run and play as hard as my older brother, my cousins, and all my friends. Bumps and bruises became a way of life for me. My limited peripheral vision meant I often failed to see obstacles at my feet, so I tripped and fell down a lot. Sometimes the consequences were much worse than bumps and bruises. At age three, I broke my leg at my grandparents' house when I jumped out of the bed of our parked pickup truck. I still remember landing squarely on the gravel driveway as my left femur snapped like a twig. A shock wave of pain coursed through my body, and I could see the horrified look on my parents' faces as they came running. I was in a body cast for the next three months, and one of my earliest and most vivid memories was the terror I felt when doctors used power tools to remove the cast. I felt certain they were going to saw off my leg. Undeterred, once the cast was gone, I started running around and playing at full speed.

A couple years later, at my other grandparents' farm, I ran headlong into a steel pipe that I failed to see in the dimming light of dusk. The warm blood ran down my face and into my eyes as my parents carried me to the car. I remember the scent of the blood-soaked towel covering my head, and that the blood left a metallic taste in my mouth as we headed to the hospital emergency room.

I was back in that same ER less than two months later, this time after splitting open my forehead on a concrete slab in our backyard. On the way to the hospital, my parents chanced upon a police officer who gave the hospital emergency room a call-ahead, and then gave us a police escort with his siren blaring. The car windows were down, and the sound of the siren terrified me. "I don't want to go to jail!" I cried as my mother tried to calm me. The ER team was waiting for us when we pulled up in front of the hospital.

By that point, I'd become such an ER regular that the hospital staff felt they had to intervene. They took me into a private room and started asking questions about what exactly had happened to me. Down the hallway, other staff were questioning my parents to make sure our stories were the same. They were concerned that I might be an abused child, but the simple truth was that my parents did not want to coddle me because of my impaired vision. They wanted me to run and play like any other healthy little boy.

Our eye doctor had recommended that I be enrolled in a local school for the blind, but my parents never considered it. Instead, they signed me up for soccer. When I was six, they put me in public elementary school, where I did fine, although there were a few bullies who made sport of my poor eyesight. A second-grade classmate liked to prank me by turning out the lights in a room, just so he could watch me struggle to find my way out. One day during music class, he taunted me with the clever nickname "blind boy." So I socked him in his clever little mouth. When the music teacher scolded me for fighting, I told her what had happened. She turned to the other boy and said, "Well, you deserved it then."

My father had been raised on a farm with his three brothers. He was an old-fashioned guy, a strict disciplinarian who believed strongly that hard work was the best way to keep young boys out of trouble. My brother and I had more household chores assigned to us than anyone we knew. Wednesdays, our father's day off, was always a big workday. We had to clean and organize everything in our two-car garage, and also restack the twenty-foot-long woodpile outside of it. We even had to wash the dinner dishes by hand, although we had a dishwasher. Twice each week, my special assignment was to clean all the bathrooms in the house from top to bottom.

One morning my dad said to me: "The yard's looking shaggy, Son. It needs to be cut."

"Sure, Dad," I said. "Do you know if we have gas and oil for the trimmer?"

"Here," he said, "use these instead." He handed me a pair of lawn shears, normally used for trimming and edging near fences and walkways. I spent that afternoon on my knees, clipping the fenced-in area of our lawn by hand.

Was my dad a little harder on me? Maybe, but I think he worried about me growing up soft, and unable to deal with the challenging life that lay ahead of me. He knew the deck would be stacked against me once I lost my sight, and that I would need an edge to get ahead. He was right. My dad's strict fathering made sure I was prepared to put in extra effort, which has been invaluable, because my everyday life requires a good bit of extra effort.

If there was any daylight remaining when our chores were done, we'd play football and Wiffle ball in the yard, or basketball in the driveway. Friends would often come by in the afternoon and wait for us to finish. That's when I first discovered my inherited talent for sales and persuasion. I managed to convince our friends that if they pitched in and helped us finish our chores, our games could start that much sooner. We often ended up with a makeshift assembly line of neighborhood kids helping with household chores.

I played organized soccer, starting at the age of four. I was pretty good and even made the all-star team a few times. Sadly, I think I was more naturally gifted at both basketball and football, but my impaired vision made it impossible for me to compete on the school team in either sport. The lights in the high school gym were too dim for me to play varsity basketball, while most varsity football games were played on Friday nights under even worse lighting conditions. Soccer was the one game always played outdoors in the bright sunlight, so it was a perfect fit for my compromised eyesight. I usually played midfielder, where I could draw upon my seemingly endless energy.

I also joined the high school wrestling team, which is when I really got into weight lifting. I was all skin and bones during my freshman year, when I weighed about 105 pounds. With daily dedication to my regular weight-lifting routine, by my sophomore year, I had bulked up to 135, and by my senior year, I was 175 with barely an ounce of fat. My metabolism in high school was off the charts. I'd go to McDonald's and drop $20 on a single meal including a Double Quarter Pounder, twenty Chicken McNuggets, a Chicken Fajita, large fries, and a couple of apple pies. I was so active all through high school—with weights, basketball, soccer, wrestling, and running—that it was impossible for me to add any pounds without the help of a special weight-gainer supplement.

At age fourteen, we went to see a top retina specialist who made casual conversation with my mom and me during my eye exam.

"So, what do you do for fun, Chad?" he asked.

"Play soccer and basketball most days," I said. "I run. I lift weights. And I'm on the wrestling team in school."

"Really?" He sounded surprised.

"Oh, and I love riding my bike, swimming, and riding motorcycles," I added. "Last summer I learned to water-ski."

"That's terrific," the doctor said. He straightened and turned to my mother. "I've examined a lot of young people with RP, and I've never seen anyone as physically active as Chad. It's very impressive, and very rare. At this age, most people with RP avoid these kinds of activities."

My teenage ego swelled with pride. In my mind, I heard the doctor say: "Chad is special. Maybe his RP isn't that bad."

After that day, I formed a cocoon of magical thinking around my condition. Because the doctor couldn't say with any certainty when I'd lose my sight, I told myself, *Maybe it won't happen until I'm much older. Maybe I'll never go blind. Maybe I'll die of old age before RP ever catches up to me.* Problem solved!

This magical thinking made me determined to live as though I were invincible. I formed a self-image apart from my eye condition. I refused to allow RP to stop me from doing anything I wanted to do. I would not be denied, and that became my identity. I would be normal with a vengeance.

I dated a few girls, rode my bicycle, and really enjoyed learning to ride jet skis and motorcycles. Despite my narrowed field of vision and difficulties with night blindness, I was a typical teenage thrill seeker with a somewhat unrealistic sense of risk and mortality.

At age sixteen, I wanted to learn to drive, and I asked my parents to sign the consent forms to qualify me for a driver's license. I got a different answer from each. My father didn't think it was a good idea, given my eye condition. My mother didn't love the idea, but she felt they had to consent anyway. She knew that my time with eyesight was running out, and this would be the only time I'd ever get to experience the freedom of driving.

They went out of their way to accommodate my driving limitations due to night blindness. I would drive to school most mornings, then drive to my after-school job working in the stockroom of a suburban clothing retailer. When I got out at 10:00 p.m., my parents were there waiting for me. I'd ride home with one of them, while the other followed behind the wheel of my car.

All along, I tried my best to ignore the periodic attacks of floaters. At first, I found I could squint my way through them, but the attacks gradually became more common and lasted longer.

The floaters would often multiply whenever I was fatigued, so I'd have to stop what I was doing and rest my eyes until they receded.

I pretended nothing was wrong, but I found myself becoming impatient and short-tempered. One day while playing a pickup game of basketball in our driveway, I missed making a game-winning shot from a spot where I almost never missed. In frustration, I turned to punch the wall between the garage doors, but instead I put my fist through a glass window in the garage door. Reflexively, I tried to pull back the punch, and in doing so, I ripped open a long bloody gash on my forearm.

"Aw, no," I said. "I think I need a towel."

"Oh my God, Chad," one of my friends said.

"You're losing a lot of blood," another said.

"Yeah," I said. "I can see that."

My friends ran inside to get my brother, and when Erick saw how much blood I'd already shed on the garage floor he began shaking visibly.

"We better get you to the hospital," he said, hurrying me toward his Ford Mustang. But as soon as we got rolling, Erick said he was afraid I might not make it in time. We drove instead to an urgent care facility about a mile from our house. The nurses and doctors kept telling me to calm down because my body was quivering. But I was already calm—it was just that my body was in shock from blood loss. I ended up with thirty-five stitches in my arm, and although I hadn't needed a transfusion, I'd lost about three pints of blood.

The doctors told me I was lucky, but I didn't feel lucky at all. I couldn't believe I'd failed to see the garage door window until my fist was already going through it.

THE DANCE OF DENIAL

After high school, I enrolled at the University of Tennessee's Knoxville campus, which wasn't far from our home. I'd always done well in science and math, so I enrolled in chemistry and biology courses, and envisioned a career in medicine. With every class, I grew more deeply invested in the idea of myself as a sighted person with a bright future as a nurse anesthetist or some other medical profession.

I was blind to the fact of my encroaching blindness. When others remarked that I seemed to be bumping into things a lot, I shrugged them off and told myself I was just having a bad day. But I had a lot of bad days. A white cane would have helped me locate obstacles at my feet and would have also signaled to others that I was visually impaired. But I never considered using a cane. Not for one moment. Instead, I walked around campus at my normal speed, which was fast, and occasionally I'd bump into some unfortunate student that I'd failed to see in my limited peripheral vision.

One sunny day as I was entering a campus building, I had an especially difficult time adjusting my vision to the dim light indoors. I was instantly plunged into darkness, so I came to a full stop to give my pupils a chance to dilate. In that moment, I felt the person walking behind me slam right into my back.

It was a slim female student. "Are you okay, man?" she asked. Neither of us were hurt, but her face wore a puzzled, faintly disgusted look. I caught the tone in her voice. It was patronizing, the way you talk to a buddy who's had too much to drink. No doubt she was thinking, *What's wrong with this guy? Is he stoned?*

11

I didn't feel I had time to explain my eyesight problems. "Yeah," I said. "I'm fine."

It was humiliating, and collisions of that sort became increasingly common as my vision continued to deteriorate. This was an especially difficult time for me, because I was ashamed that I couldn't see as well as everyone else. I simply refused to accept what my eyes were telling me. In retrospect, life was much harder for me when I had very poor eyesight than when I finally had no eyesight at all. It's easier being totally blind than being half-sighted.

During the winter months, when the sun set over Knoxville's western hills well before 5:00 p.m., I avoided driving by telling myself I'd get more exercise if I walked. The truth, which I hid from myself, was that I was scared of wrecking the car or killing a pedestrian in the dim winter light.

Even on foot, I found myself flirting with disaster. One day while walking to campus, I decided to take a shortcut through an empty house that was still under construction. I wound up falling about fifteen feet to the ground because I couldn't see that the builders had yet to install a back staircase. It was terrifying to fall that distance. This time I was lucky to land uninjured on the soft backyard grass.

In my junior year, I was eager to start living on my own, so I moved into a group house with six other friends in South Knoxville, which was closer to campus. The house was a spacious single-family home that had seen better days. There were four bedrooms on the main floor and a few more in the basement. It was furnished haphazardly and had an empty, almost haunted feeling to it.

The seven of us were not the best influences on one another. The smell of cigarette smoke filled the house most of the time, and unless we were at work or school, we'd be found just lazing around, watching sports, playing video games, or shooting pool on the billiard table I'd brought from home. Having a good time was all that really interested us in that house. It was a party house with a constant flow of visitors every weekend and most weeknights, too.

I was approaching the age of twenty-one, and the number of days I enjoyed unobstructed vision kept dwindling. The floaters became a regular source of torment and could attack at any time, whether I was tired or not. I found myself struggling to make my way through unfamiliar locations, or in any setting that wasn't bathed in bright sunlight.

My final day of driving came on a sunny afternoon on busy Cumberland Avenue in Knoxville. The sky was so bright that the glare on the windshield of my maroon Honda Accord made it impossible for me to see the roadway. The sudden fear of hitting someone or something gripped me tightly, as if someone had placed a hundred-pound weight on my chest. Stricken with panic, I braked to a sudden stop in the left lane. Then I carefully pulled the car onto the median, out of the flow of traffic. Covered in a cold sweat, my legs were shaking from the adrenaline that pulsed through my body. I phoned a friend to come get me, and we left the car sitting there until we could get someone to go back with us to pick it up. My beloved Honda Accord went up for sale soon after, and I never dared to get behind the wheel again.

Looking back, it's hard to explain how I continued to deny the seriousness of my condition after that day. The world of sight kept fading away from me. With each passing day, reading became just a little more difficult. I maxed out the light-bulb wattage in my desk lamp. Then I learned to switch off my room's overhead lights when I was studying, just so I could enjoy a little extra contrast on the pages of my reading materials.

Until that fateful evening when I could not read my literature assignment, I really had no idea how deeply ingrained my self-concept as a sighted person was. In my studies, I'd come to realize that I was naturally a *visual* learner. I had such a strong photographic memory that while taking a test I could picture in my mind's eye the pages in my textbook where the answers appeared. Nearly everything I knew and liked about myself was related to visual experiences—whether it was sports, video games, watching movies, or taking photographs.

Now it was clear that soon my eyesight would be gone forever. In that moment, my world collapsed. My dance of denial was over. I was going to lose my autonomy, my independence, my future—my everything. I would never have a career in medicine. I'd be lucky to have any kind of career at all. And if I should ever marry and have a family, I'd never get to look into my wife's eyes, or see the smiles on my children's faces. Everything I knew and everything I'd hoped for vanished as if I'd just awakened from a wonderfully vivid dream. In its place was my new dark reality marked by disability, dependency, and low expectations.

I sank into a pit of despair in the months that followed. I was so run down with self-pity that I could only imagine the darkest prospects for the future. Soon I would be turning twenty-one, without a college degree, living off government checks and listening to audiobooks in the same room I'd grown up in. I was in mourning. I was grieving the death of an imagined future self. For almost three years, I'd been working hard toward a career in medicine so I could help people. Now not only was that dream gone, but I doubted whether I could even help myself.

If I had one refuge during those miserable months, it was the weight room at the gym. I went there more frequently than ever. I upped the intensity of my workouts as an outlet for my overwhelming feelings of frustration and pain. Maybe I unconsciously thought I could somehow exercise my way out of my predicament—who knows?

I did know that the prospect of blindness felt completely overwhelming. On one hand, I feared I didn't have what it takes to overcome all the obstacles to living a good life without eyesight. On the other hand, I was ashamed to feel afraid. It felt unseemly—and downright unmanly—to allow my new circumstances to get me down. But I was embarrassed to be blind, with all its implications of weakness and dependency. I was ashamed that it had happened to me, even though I knew it wasn't my fault. It wasn't anyone's fault.

I continued to go to classes that fall and hoped to finish out the semester. The futility of my situation didn't really hit me until I attended an anatomy class featuring a human cadaver, a pale, older female corpse sliced open on a table in the middle of the room. The professor was pointing out features of the internal organs, but I couldn't see well enough to grasp what he was saying. I moved closer, but to no avail. Unless I was ready to feel my way around the cadaver, that day's lesson was lost to me.

I sat there in a funk for the rest of the class. My hopes and dreams for my future self felt as cold and lifeless as the cadaver laid on the table in front of me.

BREAKING THROUGH

I self-medicated my negative thoughts often during these months. I partied and hit the booze harder than I ever had before. My personality shifted, and I became moody and willful. One night, after a few drinks too many, I decided I wanted to go to Krystal—a local burger chain similar to White Castle.

Unable to drive, I asked my girlfriend to take me, but she refused. I was determined to walk there on my own, even though my vision at this point was reduced to making out large shapes and changes in light. I figured I could find my way to Krystal by navigating from one streetlight to the next.

Even if I'd been sober, this would have been a foolhardy mission. With more than two miles still to go, I tumbled about ten feet down into a roadside drainage ditch. My legs were in so much pain, I could barely drag myself back up to the road's edge.

Just as I was finally getting to my feet, a car pulled to a stop in front of me. I squinted at the headlights and hoped it wasn't a police car.

"Chad? Are you all right?" It was my girlfriend's voice. She'd come looking for me. "Chad, what are you doing?"

I snapped, "I'm trying to get some Krystal burgers." I was filthy and limping.

"You're a mess. You can't even walk."

It was obvious I'd badly injured my legs. She drove me to my parents' home, and my brother took me to the hospital. It was 2:00 a.m. when we arrived. By the time we left at 4:00 a.m., I was on crutches. The next few months were especially difficult, hobbling around on crutches when I could barely see where I was going. I was used to walking fast, and now I could barely walk at all.

I became a glutton for anything that salved my constant internal ache of dread and shame. I took solace that I had a group of housemates who seemed to accept me as I was. Sure, I was demotivated and cavalier toward my future, but so were most of them. My brother cautioned me that moping and getting drunk all the time wasn't like me. He wanted me to get my act together, but I wouldn't listen.

Then my older cousin Mark came to visit from Houston. Mark and his younger brother, Jeremy, had been best friends with my brother and me when we were all little kids. But after his father, our uncle Paul, got a company transfer to Texas, we didn't see one another for about ten years. Nonetheless, the bond remained between us, and I looked up to Mark as my cool older cousin who had it together and was making a life for himself in Houston. I also looked up to him literally, because he was a big, tall guy, at least six foot four.

On the evening Mark came to see me, we went out for dinner and a few beers on The Strip, Cumberland Avenue's commercial district near the UT campus. Then we went back to my place on Woodlawn Pike, and he met many of my housemates. After a while, he asked me to come outside with him. We stood in the gravel driveway, leaning against the back of one of the parked cars.

"Chad, you know those guys in there?" Mark asked me, gesturing toward the house.

"Yeah?" I didn't know what he was getting at.

"All those guys are losers."

It was a blow to hear him say it, but I knew what he meant. It was nothing I hadn't heard before from my brother.

"None of them are going to amount to anything in life," Mark said.

All I could do was look down at the gravel driveway. My ears started burning with shame to hear Mark's judgment.

"And if you keep hanging around them," he continued, "you're going to be a loser, too."

I remember the look on his face when he told me this. I glanced down at the gravel driveway and kicked at the rocks with anger and frustration. This was Mark's message of tough love. He cared too much about me to try and cheer me up with happy talk. Is it better to tell someone what they want to hear or the hard truth that they need to hear? We didn't say much to each other for the rest of the evening because we didn't need to.

After that day, my eyesight continued its decline, but the fog in my head began to clear. I didn't leave my housemates, and I didn't give up drinking, but I thought of Mark's words often.

What I was contending with, I realized, was hopelessness. Everything worth doing in life begins with the hope that success is possible, and there will be a better tomorrow. But I felt hopeless because I was completely ignorant about what was possible for me as a blind person. I had not prepared myself for the road ahead of me. I hadn't learned to read braille or use a cane. I hadn't identified any support or resources to help me. I'd rarely bothered to visit the disability center at UT. I had no idea what success would look like as a blind person. My overwhelming hopelessness was the price I paid for my years of denial, years of pretending I had a future as a sighted person, years of ignoring what my own eyes were telling me, that my life as a sighted person was on borrowed time.

Today I know that whenever we indulge our feelings of hopelessness, we're just signing up for a life that conforms to our worst fears. But fears about the future have no basis in reality. Twelve-step programs refer to fear as "False Evidence Appearing Real." In my experience, that false evidence melts away whenever we choose to confront the reality that appears so threatening.

But here is where blindness becomes a wild card in the equation, as does any disability or unusual circumstance: it gave me an easy excuse to never try. All the hopes I once had for myself died the day I accepted I would soon lose my sight. No one would ever blame me for feeling heartbroken and hopeless after that. The world will always be willing to accept whatever excuses I have to offer as a blind person. I will always have a socially acceptable reason to underperform and disappoint. So what if I'm late, or show up unprepared? Cut me some slack. Give me a break. I'm blind, you know?

Making excuses, though, had never been in my nature. My whole life I'd been testing the limits of what I could do with my deteriorating vision. Now that I'd gone blind, why throw in the towel? My eyesight was gone, but my hunger for life was unchanged. There were so many things I still wanted to do with my life. I felt it in my body, as sure as the pull of gravity. There was a fire of life that still smoldered deep inside me. If I could, maybe I could use that fire to beat back my fear of the darkness closing in on me.

With the start of a new semester at UT, I began taking business courses, because I thought that a business administration degree would offer me the widest range of career options. But my change of direction came at a tremendous cost. At least eighty-five credit hours I'd earned in the sciences would go completely to waste, because none of them could be applied to my new major. I'd have to do the previous three years of college all over again. It felt like a punch in the stomach to see all those hard-earned credits and years of work go down the drain, all because I had failed to come to terms with my fading eyesight.

And yet, changing majors was a simple fix compared to my larger problem of no longer being able to read the printed word. Here I was, a natural visual learner who had lost his ability to see.

I would have to relearn how to learn—literally. Back in 1998, there were relatively few technologies to assist the visually impaired. Audiobooks were rare, and none were available for any of my assigned readings.

My mother came to the rescue at this point and performed the Herculean task of reading aloud all my assigned texts onto cassette tapes. Every evening after coming home from work, she sat down at the kitchen table and began reading aloud into the recorder. With every spare moment she had, sometimes late into the night, she committed to audiotape everything I needed to read that semester. She read thousands of pages, night after night, until her voice wore down to a rasp.

In concert with my mother's textbook readings, I started recording all my lecture classes onto microcassettes, and the university provided me with a notetaker to go over my lecture notes with me. With time, I developed a new learning system for myself, where I would listen to each lecture and recorded reading assignment at least twice.

I never learned to use a cane. I decided I wanted a guide dog instead. From what I'd gathered, a guide dog would be essential to the kind of independent life that I wanted for myself. I knew that, after college, it was possible that I would move to another city, perhaps one that was far away from my family in Knoxville. Almost all the jobs I wanted involved a lot of business travel. A guide dog could be my trusted companion and my ambassador to that world.

My parents discovered that a local Lions Club chapter in Knoxville maintained a special partnership with a Michigan training center called Leader Dogs for the Blind. In an amazing gesture of kindness and generosity, the Lions Club agreed to sponsor my training, along with the substantial expense of acquiring a highly trained guide dog. My friends and family members helped me record videos required by the application to Leader Dogs. To match me with the right dog, the trainers there needed to see how much I walked each day, how fast I walked, the length of my strides, and the variety of street crossings and other obstacles I typically encountered.

Not long after I moved in with my housemates in South Knoxville, my parents moved to a new house they had built a few miles away. The house had been under construction for about a year, and its custom floorplan owed a lot to my mother's influence. When I went to visit, I noticed that they had outfitted the basement as a two-thousand-square-foot apartment, with a full bath and a private entrance that led directly to the back patio. A section along one wall had plumbing and electrical fixtures that would accommodate a kitchen installation.

I knew what they'd done, although we didn't discuss it at the time. They'd built me a private apartment in case I ever needed a free place to live. If the world ever got too tough for their disabled son, they could offer me a soft, comfortable place to land. They went to this expense and trouble out of their profound love for me. And although their intent was squarely focused on helping me, I wanted no part of it.

In my mind, I called that place the "Loser Basement." I was determined to never spend a day in it as long as I lived.

TWO

HAPPINESS
IS NOT A FEELING

"I'M GONNA GO DOWN AND CHECK MY EMAIL."

Is this some kind of joke? That's what I remember thinking.

The voice belonged to Steve, my roommate during my month-long guide dog training at Leader Dogs for the Blind. At that moment, I was certain Steve must be kidding around, because I assumed checking email was one of many things blind people are unable to do. In my mind, he might as well have said, "I'm gonna go outside to play a pickup game of basketball. Want to come watch me dunk like Michael Jordan?"

So I kidded him right back.

"Yeah, right," I said with a laugh. "Good luck with that."

"No, I'm serious," Steve said. "Down in the library, they've got a JAWS system for reading email. You should try it."

JAWS stands for "job access with speech." It's a computer application that enables blind people to operate Microsoft Windows through keyboard shortcuts and spoken feedback. For most blind people who use a computer for school or work, JAWS is an absolute necessity. The fact that I'd never even *heard* the term *JAWS* until that day shows how ill-prepared I was to function without sight.

The campus of Leader Dogs for the Blind is a beautiful facility in Rochester Hills, Michigan, just north of Detroit. When I arrived there in the summer of 1999 for my month of guide dog training, I wasn't in the best head space. I was coping with a nagging sense of dread about the future. I'd spent my whole life up until then behaving like I could see fine. Using a guide dog would blow my cover. I'd be parading around with my faithful canine friend by my side, openly and perpetually declaring my blindness to the world. I'd have to embrace the very thing I'd been avoiding all my life, and it made me very uneasy.

It was a difficult period of adjustment, compounded by the way I kept forgetting I was blind. That probably sounds crazy, but I visualized my surroundings so vividly that I would often step forward as if I could really see—and bang into a desk or chair I'd failed to visualize. It didn't help that I was accustomed to moving fast whenever I decided I had something to do. Dozens of times each day, I'd reflexively start moving and then have to slam on the brakes, because all I could see were vague shifting shadows and shapes. It was like reliving the trauma of loss, over and over again every day.

I'd felt relieved when I was accepted into the Leader Dogs program, and the local Lions Club members had assured me that other blind people whose guide dogs they'd sponsored were overjoyed with how much their lives were improved. I hoped they were right, but I wasn't at peace at all. I was always fighting back the clouds of sadness and self-pity that hung over my head.

There was also some degree of uncertainty about the weeks ahead. A small percentage of students at Leader Dogs wash out of the program, either because they can't master dog-handling, or because the necessary bond between student and dog never takes hold. In cases like that, the student gets sent home without a dog, and although I never considered failure as a possibility, I also didn't know nearly enough to be 100 percent sure.

A GLIMMER OF GRATITUDE

I checked into my dorm room, where I was matched with my roommate, Steve, who was a smoker, like myself. I did my share of smoking and drinking in the dark days before I showed up at Leader Dogs, but there was no alcohol permitted on the campus. That gave me a little bit of a problem, because I didn't really enjoy smoking without a drink in my hand. Smoking and drinking were like peanut butter and jelly to me. So I decided if I couldn't drink, I wouldn't smoke, either. I'd be totally on the wagon for twenty-six days—probably the longest stretch I'd gone without drinking or smoking since high school. (As it happened, I ended up quitting smoking during my days at Leader Dogs.)

My roommate, Steve, was about twenty years older than me. He'd gone through all his Leader Dogs training many years earlier. Now he was back on campus to train with a new dog after retiring a dog he'd had for about ten years. He filled me in on how the program worked, and which trainers he'd worked with in the past. I learned that Steve had been blind from birth. Unlike me, he'd spent every day of his life navigating through a world he'd never seen.

Talking with Steve about JAWS and other things reminded me how rarely I had interacted with other blind people before visiting Leader Dogs. I'd always avoided any experiences that might remind me of what the future had in store for me. There are people with aversions to hospitals and funerals because they dislike being reminded of their mortality. That's how I had been with my impending blindness. Although Steve had offered to teach me how to use JAWS during my month at Leader Dogs, I never took him up on it. I wasn't in the right frame of mind for it at the time.

Training at Leader Dogs is intense. It has to be because the time there is limited, and there's a lot of ground to cover. Leader Dogs runs a dozen monthlong classes each year, with no more than twenty-four people in each class. Inside each month (just twenty-six days, really), each student needs a crash course in dog psychology and behavior, plus intensive training in orientation and movement—how to navigate the environment and give instructions to the dog in ways the dog will understand. The days are long, and working with the dogs requires repetition until the dog and blind handler team are performing comfortably enough to venture out into the world.

Building a tight relationship between handler and guide dog is essential because the two of them will need to seamlessly and safely navigate the world without any assistance from others. The stakes could hardly be higher. One single mistake or momentary lapse by handler or guide dog could get both of them killed.

The first few days of Leader Dogs training were devoted to what's called *Juno training*. I was shown how to hold the harness and give the proper commands to a fictional dog named Juno, while my trainer held the other end of the harness and mimicked the action of the dog—pulling, stopping, and even misbehaving when I made a mistake or gave an unclear command.

On the morning of our third day there, all the students were assigned their dogs. Rich, one of the professional dog trainers at Leader Dogs, introduced me to my new constant companion, a handsome seventy-five-pound German shepherd puppy named Miles. Rich told me to take Miles back to my room for about two hours, until lunch. The idea was to spend time alone with Miles, to begin the process of bonding with him. As I led Miles away, he kept craning his neck to look back at Rich. A guide dog can be loyal to only one handler at a time, and for Miles, Rich was his master. It would be my job for the next several days to get Miles to "turn over," and transfer his loyalty from Rich to me.

Back in my room, I got down on the floor with Miles and offered him constant attention for two solid hours. I hugged him, ran my fingers through his fur, and told him what a good boy he was. He was truly a sturdy, powerful beast, even though he wasn't yet full grown. I'd specifically requested to be assigned a German shepherd in my application, because I liked to be on the go, and wanted a dog that also enjoyed being on the move. And German shepherds are known for having an extremely high drive for work. In my application video to Leader Dogs, the trainers saw that I was fit, athletic, and liked to walk fast. At six feet and almost two hundred pounds at the time, I was able to handle any of the larger dog breeds. And if you need someone to guide and watch over you, what could be better than having your own shepherd?

Miles seemed anxious to be separated from Rich and kept bobbing and weaving for the door. I grabbed the leash and gently but firmly led him back. This made me a little uneasy because I didn't want Miles to think I was a bully, who was just keeping him from his handler. I'd learned in our lectures on dog psychology that the start of the transition period is an extremely tenuous time. If Miles thought I was coming across too harshly, he could decide that I was a hopeless jerk, permanently fracturing our relationship. And if that happened, I could be going home without a dog.

So I needed to turn myself over to Miles in order to get Miles to turn himself over to me. On the other hand, if I failed to take the lead in the relationship, Miles might decide that he was the alpha in the relationship and wouldn't take my commands. Miles needed to both love me and respect me. It's a delicate balancing act.

"You want Miles to believe you're the coolest guy this side of the sun," Rich said. At that point, when that bond had formed between us, Miles would gladly obey my commands, learn new commands from me, and be ready to do anything I wanted from him. Dogs are pack animals that crave structure. Having steady, consistent leadership makes them feel happy and secure.

Lunchtime that day was loud, joyous, and chaotic. Everyone was giddy with excitement, chattering about their new "special partner." "His fur is so soft!" and "Look how happy he is!" "He can't stop wagging his tail!"

After lunch, we walked around indoors, giving the dogs practice in heeling at our sides. When it was time to go outdoors, we all lined up and were instructed to take a plastic bag from the dispenser.

"Plastic bag, what's this?" I said.

"Time to take the dogs outside to do their business."

I paused for a moment, and joked, "Hey, this was not in the brochure."

"Well, Chad, who did you think was going to pick up after your dog?"

Until that moment, I hadn't given it any thought. It didn't occur to me that having a guide dog meant I'd be picking up dog poop in plastic bags for the rest of my life, but it is a small price to pay to have my best friend at my side at all times, and be able to walk wherever I want to go.

My main problem with Miles during my first few days with him was that he was always looking for Rich, because he still had that bond with him. So here I was, trying to train Miles to obey my commands, and Miles couldn't take his eyes off of Rich. When Miles would twist his head, I had a hard time telling if he was sizing up an obstacle for us to go around, or if he was just looking at Rich, again.

What I discovered is that German shepherds are very loyal by nature and as guide dogs they take longer to "turn over" from one handler to another. Labs and other retriever breeds may turn over in a few days, while German shepherds could take a week or more.

One day during that first week, Miles was off leash in my room with me, when my roommate entered unexpectedly. Miles bolted past him out the door and disappeared down the hallway. I made my way down to Rich's office and, before I could get a word out, I heard Rich laughing.

"Looking for someone?" he asked. Miles had made a beeline to his office and was sitting loyally by Rich's side.

For the next three weeks of our training, we'd get into shuttle vans each day and go to downtown Rochester where we would practice "curb work"—the fundamental skill of walking down the street with the guide dog and stopping at each curb and obstacle, while the dog awaits the next command. Like all the dogs, Miles had been taught all the guide dog basics before being assigned a blind handler. He was comfortable wearing the harness and learning to take cues from me. He knew left from right, and to stop at curbs and stairs. He'd learned not to run me into obstacles, and to always seek the path of least resistance once he'd been shown the direction to go in.

It was during these first few days downtown that I could feel the trainers' promises coming true: my life really was about to change. Walking with Miles transported me into another world, one in which strangers were almost always kind and welcoming. It was a world apart from where I'd been living for years, where I was the clumsy stranger to be avoided, assumed to be intoxicated because of the unsteady way I walked and bumped into things. Now I was striding confidently and independently inside a bubble of positive energy, as though I had an aura around me. In the eyes of strangers, I was no longer a klutz. I was a fit and handsome young blind dude walking with a confident stride next to the cutest and smartest dog on the planet.

As great as it was to be walking with Miles, much of the time spent during those daily trips downtown could be very tedious. One trainer was assigned to supervise a group of four students and their dogs. So, while that trainer was out doing curb work with one student, the rest of us were just sitting there with our dogs, waiting our turn.

At first, I found myself getting impatient with my fellow students. I'm a fast learner, and at that time, I was quick to judge people who couldn't keep up. It seemed to me that some of the students had a tendency to resist the fundamentals of dog psychology. They were reluctant to correct the dog when he didn't perform the commands, so I heard them try to reason with the dog, as they might with a child. Or, I'd overhear them give conflicting commands, sometimes three or four at once. I wondered, what was wrong with these people? Weren't they paying attention? If the dog can't clearly hear the instructions, he could get confused and put the handler in danger of tripping, falling, or worse.

As the first week passed, however, I got to know the other people in my group a little better. And although I'd gone to Leader Dogs for guide dog training, I ended up learning a far more valuable and impactful lesson from my fellow students.

It was a lesson in *gratitude*.

Many of my fellow students had serious health problems associated with their blindness. Two of them were so physically compromised, they had to go out for kidney dialysis during the week. Others had physical and mental impairments that affected their sense of direction and learning abilities. I overheard one woman give her dog five directions at once, and I huffed silently in frustration. Then I heard her trainer slowly and patiently prompt her to try again, to give just one command at a time. Eventually, she got it. I discovered later that she had a cognitive disability. Like a complete jackass, I'd been silently judging her for not paying attention.

Then there were the two young women who were both deaf and blind from birth. They had an interpreter with them who communicated by tapping code into their hands, similar to the way Helen Keller had learned to communicate. As I conversed with these young women, through their interpreter, I was deeply moved by their courage. I tried to imagine not being able to hear, and the prospect of that isolation shook me to my core.

It was hard for me to imagine what it would be like if I also didn't have my hearing. When I cross the street, I rely on sounds to orient me to the flow of traffic, which allows me to cross more safely. I can hear voices and footsteps approaching, so I'm not startled when someone brushes by me. And I can always ask for directions. I realized how much I'd prefer to be blind than deaf, because my entire social existence depends on listening and speaking.

The great courage these amazing people demonstrated daily made me ashamed of my own self-indulgent negativity and toxic self-pity. It was as though all my good fortune, all the positive things in my life, had reached up and smacked me square in the face. For the first time in a long time, I counted my many blessings. I'd enjoyed almost twenty-one years of eyesight, with all the pleasures and experiences it granted me. I was fit and strong, I had good hearing, a sharp mind, and overall good health. I was socially confident and had an active dating life. Above all, I had a loving, supportive family.

Most of my fellow students had very few of these things, and a few of them didn't have any at all. For some of them, the chief benefit of having a guide dog would be that they could go to the store by themselves for the first time in their lives. They were every bit as hungry for life as I was, and that modest vision of happiness and independence filled them with joy. What was my excuse for feeling anything less?

In the course of four weeks at Leader Dogs, the everyday bravery of those two deaf and blind girls and my other fellow students had taught me a profound lesson in gratitude. It was one that I don't think I could have learned any other way.

By the time I returned to Knoxville with Miles, I felt I'd lost my right to complain about losing my eyesight. My mindset hit a tipping point, where now I got annoyed and impatient with myself whenever I felt pangs of self-pity. How could I ever dare feel sorry for myself with any semblance of self-respect? It was high time that I be grateful for the cards I'd been dealt in life and play them with everything I had.

MILES THE MAGNET

Miles's new home in Knoxville was a two-bedroom apartment near the UT Knoxville campus, where I'd moved in the spring semester of 1999. Within the first hour of introducing him to the apartment, I failed a major test in dog psychology.

I'd been told that Miles needed to be leashed in the apartment, attached to a tie-down of some sort, ideally bolted into the baseboard. I didn't have the chance to get that done before I left for Leader Dogs. Now that I was back home, I wanted to take a shower before going out that Saturday night. Tentatively, I tied Miles's leash to the oven door in the kitchen and headed for the shower that was just six feet around the corner.

Fifteen minutes later, when I emerged from the bathroom, I was alarmed to discover that Miles was missing from the kitchen. Also missing was the oven door. I called out Miles's name and walked around the apartment until I found him, curled up by the front door, with the thirty-pound oven door hanging from his neck.

My mistake was that I'd tied up Miles in a spot where he didn't have a clear view of the apartment. When German shepherds are left alone, it turns out, they have an instinctive need to see the entry points of the room they're in. The kitchen counter blocked his line of sight to the apartment door, and Miles decided he needed to escape the kitchen. The oven door hinges were no match for Miles when he was in beast mode.

Like most service dogs, Miles had been raised from his earliest days for the role. As a puppy he'd been conditioned to behave himself in all kinds of distracting environments—public transit, department stores, grocery stores, and restaurants. He'd likely been taken to the dog toys section in Walmart or Target, to ensure he wouldn't be distracted when toys tempted his attention.

Miles had also been desensitized to pungent food smells and noisy environments that would drive other dogs to distraction. If I wanted to go to a steakhouse for dinner, Miles would be able to guide me to my seat and then lie still beside me for the entire dinner, completely ignoring the waiters passing by with trays of sizzling meat. Loud, crowded bars, where I often enjoyed going, would have no effect on Miles. He'd just wait quietly at my feet until I was ready to leave.

Leader Dogs often had alumni come in to talk to us, and for one session there was a kindly fiftyish southern gentleman, a businessman named Mike, who'd had several dogs trained by Leader Dogs over the years. After he finished his talk, he and I were conversing casually, and he got to know me a little. He took me aside, as we continued to chat in a warm conversation.

"Son, things are really going to change for you," he said.

"What do you mean?"

"Your social life is going to change. It's going to be good."

I tried to be agreeable, but I wasn't sure what he was getting at.

"Really, what do you mean?" I asked him.

"Let me ask you something, son," Mike said. "How many times has a young lady come up and asked if she could pet your *cane*?"

I laughed. "Zero." I didn't bother to tell him I didn't even have a cane.

"Trust me," Mike drawled. "It's gonna be good."

And boy was he right.

Thanks to Miles, I spent the next few years in Knoxville rarely having to pay for my own drinks. I can still remember all the *oohs* and *aahs* the first time I entered one of my favorite watering holes with Miles by my side. Young women would flock to Miles, stroke his fur, and coo in his face. They'd insist on buying the drinks. From there, shifting the conversation from Miles to me was a breeze.

One evening, my friends and I were drinking at one of our favorite spots in Knoxville's Old City area. A young lady came up to greet Miles, bought us both drinks, and after we'd conversed for a while, she wrote her phone number on a paper bar napkin, which I put in my pocket. About an hour later, another young lady I'll call Cathy approached and made the same fuss over Miles before buying us drinks. When Cathy offered to give me her phone number, I had to apologize.

"I don't have anything to write on," I said. "I have a napkin in my pocket, but it's got someone else's number on it already!"

"That's no problem," Cathy said. "Give me the napkin. I'll just write my name and number on the other side."

My friend Corey had heard the whole conversation, and after Cathy had moved on, he came up and grabbed my arm.

"Two women's numbers on the same napkin?" he exclaimed. "Come on, Foster, that's excessive."

"Either you got it or you don't," I told him.

"What you've got is Miles," Corey said. "Why don't you just stay here and let me walk around with him for a while?" I knew he was only kidding. I don't think Miles would have budged an inch if Corey had tried pulling on his harness.

My friend Ken chimed in, "Yeah, Foster, you're only getting that attention because of the dog."

"Maybe," I said, shrugging. "There aren't many advantages to going blind, so you better believe I'll take the few there are."

Corey said that I should at least be honest about what I was using Miles for, by changing his name, from Miles to "Magnet."

Sometime later, when I was in San Diego visiting Ken, a group of several young guys struck up a friendly conversation with us. It took us a while to figure out what was going on. They'd noticed Miles was attracting women in our direction, so they decided to join our conversation knowing the women would assume they were in the same group as Miles. Hats off to those guys for such a brilliant move.

During my Leader Dogs training, I'd been cautioned that it's a bad idea to let other people pet my guide dog. For the handler–dog partnership to work, I needed to be *everything* to Miles, his sole source of attention, affection, and, if he misbehaved, correction. Miles has to believe that I've hung the moon and stars in his world. That way he's certain to keep his focus purely on what I tell him to do. Dogs are terrible multitaskers, even worse than humans.

So, I discouraged most people from petting Miles. But because I was a young man with a healthy interest in young women, for them I occasionally made an exception.

Then, one day, Miles and I almost paid the ultimate price for breaking the no-petting rule. We were walking to class on the sidewalk along Forest Avenue in Knoxville, when Miles suddenly bolted into the street, pulling me behind him with one hand on the harness and the other tightly clutching his leash. To my left, I could hear the honking of a bus horn and the squealing of its brakes. Then, as Miles took us farther across the street, I heard car horns blaring and tires screeching to a halt on my right.

At the curb on the other side of Forest Avenue was a young woman who lived in my apartment building. She always made a big deal over Miles whenever we met in the hallways. Miles had seen her standing there across the street, and his dog brain lit up: "Attention!" In an instant, that urge to be stroked and hugged overrode his years of conditioned training. He nearly got us both killed.

I was really shaken up by the experience. Miles had never done anything like that before and I gave him a strong verbal correction to etch this incident into his memory as bad behavior. That was the last day anyone ever got to pet Miles while in a harness. I put a sign on the harness, which you now see regularly on guide dog harnesses: "Please don't pet me. I am working."

Miles still remained a people magnet and a great icebreaker for conversations in social situations. But I had to explain to people why they couldn't touch him. If anyone ever persisted, I'd tell them about the day Miles dragged us in front of a bus just to get a hug from a friend in our building. That was usually enough to persuade them that they should look and not touch.

CHOOSING HAPPINESS

With Miles, my everyday life changed just as the Leader Dogs people had promised. Navigating the campus felt like I'd entered a different world, even though all the places were very familiar to me. I was now moving inside that same aura of positivity and good vibes that had surrounded me during my training with Miles on the streets of Rochester.

Wherever I went, I found I could easily ask for directions and assistance because Miles was a handsome symbol that I was disabled. In a mall, or some other place that I wasn't totally familiar with, I found that when I stood still, people would approach and say: "I love your dog. Can I help you find your way?" Miles was the ultimate goodwill ambassador. Everyone loved him at first sight and wanted to be of help to his handler.

Walking with a guide dog is a process of point-to-point navigation. Miles could lead me to an escalator when I said "escalator," but Miles wouldn't know which escalator I needed to take, except when a particular escalator was part of our routine. In unfamiliar places, I'd normally ask for directions and then use my orientation and mobility skills to give Miles specific instructions about which way to go.

That's how we functioned as a team. It was up to Miles to follow my commands while guiding us around people and other obstacles along our path. For a blind person, that's a completely different experience than walking with a cane, where the purpose of the cane is to orient the user and probe for obstacles ahead. If I'd ever learned to use a cane, guide dog training might have required a period of adjustment, to surrender the feel of objects in front of me and suddenly start gliding through an obstacle-free world.

Miles wasn't perfect and neither was I. He probably needed more practice spotting overhead obstacles than I had given him, and one day he led me full stride into a horizontal steel rail, part of the scaffolding at a construction site on campus. It had been set too low and caught me right in the forehead. I recall lying semiconscious on the sidewalk, but for how long I can't say. On another occasion, I was leaving UT's Student Disability Services and walked in full stride up to what I thought was a curb, when I felt Miles come to a full stop. As I gave the "forward" command, I made the cardinal error of taking the first step instead of waiting for Miles. But Miles had halted because he could see we were standing at the edge of a high retaining wall. I never let go of Miles and we both took a ten-foot tumble to the ground below. As I stepped forward and felt only air beneath my foot, I remember thinking, *Okay, Miles, we're going for a ride.*

As the fall semester at UT began, my revelation at Leader Dogs about gratitude had an enormous impact on me. Day after day, I'd thought about those deaf and blind girls, and how joyous they were to gain some independence within their dark and silent world. I even experimented occasionally by wearing headphones for short distances, just to see what it would be like to be walking without the benefit of both sight and sound. Even though I knew I was totally safe with Miles as my guide dog, the experience terrified me.

The example of those two girls forever transformed my outlook on life. I'd returned to Knoxville with Miles, but also with a deep, deep sense of gratitude that permanently shifted my attitude. It was the start of a new chapter for me, with a new perspective that I've lived every day since.

I knew for sure that I'd changed because people on campus were treating me differently. Doors started flying open for me. Professors were more willing to be helpful, to go the extra mile for me, more willing to make accommodations for exams. They could see how I was tackling my problems with such a positive attitude. I made it easy for them to meet me halfway.

I ran into a significant obstacle in one of my finance classes because it dealt with mathematical formulas for assessing risk and return. Classroom exercises and discussion revolved around pointing out calculations up on the whiteboard, making it almost impossible for me to follow along.

The teacher, Suzan Murphy, had a reputation for being tough and making all students earn their grade. When I mentioned the problems I was having, she asked me to come back later in the day.

"We'll come up with a plan," she said.

We met at 6:00 p.m. and went over the specific formulas I was having the most trouble with until I fully understood what had been covered that day. After meeting with her several times, she offered to help me find a tutor to give me grounding in the basics, and then I could check in with her every week to make sure I wasn't missing anything.

It was an unusual arrangement, but so was my situation. I knew that it was rare to get the level of help she provided me, but I also knew that I had earned her respect. I'd told her I was striving for straight A's that semester, and I'd demonstrated with my preparedness that I was serious about that goal. I think that's what inspired her to make so much extra time for me. And, unlike the vast majority of students in Suzan Murphy's finance class that semester, I got an A.

After those twenty-six days at Leader Dogs, I had stopped dwelling on things I couldn't change. Now all I wanted to do was focus my energy and efforts on the things that I could change. People like Suzan Murphy picked up on the energy I was giving out, and they were more willing to contribute their own energy to help me.

I contacted a state social services agency, Vocational Rehabilitation Services for the Blind and Visually Impaired, to ask for JAWS training and the software that goes with it. Larry, my counselor at the agency, wasn't very glad to hear from me. He'd been turned off by my poor attitude in the past and had raised concerns about my grades. So, he didn't want to give me the JAWS software.

"Why don't we start with a training session first to see if it makes sense," Larry said. "Then, if you like it and your grades are on par, we can talk about getting you the software, and maybe even a computer. How does that sound?"

It sounds like garbage, I said to myself. I didn't like being toyed with, and Larry seemed to be holding the software over my head until I could prove I was worthy. In the past, I'd hold a grudge, even though I wouldn't be hurting anyone but myself. Now, I had to acknowledge that I'd earned his skepticism. It was time to step up and earn his respect.

The JAWS computer training was extremely difficult, much harder than I had imagined. The system converts words and symbols on the computer screen into synthesized speech output I could hear through earphones. Not only was the voice synthesizer hard to understand, but there was an entire JAWS taxonomy I needed to learn—more than a hundred different control terms that helped translate visual data into auditory information. JAWS also worked one way on the web and a different way on traditional Windows applications, so users also have to remember which environment they're working in. With certain computer applications, JAWS didn't work at all. I could get stuck wondering: *Is JAWS not delivering the information, or is the information simply not there?*

Over the next several months, I learned JAWS and overcame Larry's skepticism. I received my copy of JAWS software and used it to help me earn straight A's in my business courses that semester, landing me on the Dean's List for the very first time. Next, I set a goal to graduate by December 2000, and took a very heavy course load all through the spring and summer sessions that year.

Throughout that time, there were occasions of setbacks, frustration, and despair. Did I have a chip on my shoulder? Something to prove? You bet I did. But I consciously chose to stop bemoaning my condition and instead start considering what resources were at my disposal. Whenever I was able to ask for help, I found it readily available.

I could not correct my vision, but I could control how I chose to see the world. I still had my cognitive faculties, superb hearing, a tremendous family, and very good health. I also happened to be living at the dawn of the Information Age, the greatest time in history to go blind.

In my final year at UT, I was able to write a thirty-page research paper for my international business class without ever going to the library. Even back in 2000, all the texts I needed to research had been digitized so I completed all the research and writing from the comfort of my apartment. I got an A on the paper, which counted toward the majority of my grade for the course.

Twenty or thirty years earlier, what would I have done? I would have likely needed a partner, assuming I could find one, to go to the library with me and propose possible source materials, reading the content to me so I could decide what I wanted to use. The actual writing of the paper, copying quotations and citations—it's all too difficult to imagine. Obviously, there were many blind college students long before there was ever an internet. But they were the exceptions. Who knows how many young blind people never even attempted to go to college because it was too hard or too expensive to hire assistants and tutors to help them get their schoolwork done.

I was lucky to be born when I was. Even back in 2000, I was the beneficiary of an emerging digital world that had just begun to level the field for people with all kinds of disabilities. The internet was lifting off and paving the way for a service economy and an information revolution never before seen. Computers and smartphones would soon saturate daily life. The prevalence of guide dogs, coupled with the vast number of technological tools at our disposal, would provide me many options for professional fulfillment, pastimes, and running my household. And right now there are dozens of ongoing clinical trials involving gene therapies and other potential treatments for retinitis pigmentosa. Thanks to organizations like the Foundation Fighting Blindness (founded by Gordon Gund, a philanthropist business leader who lost his sight to RP at the age of thirty), I might someday be able to see again.

Out of all this, I came to realize that happiness isn't a feeling. It's a choice we make. In the years that would follow, I could look back at the feeling of gratitude that came to me during Leader Dogs training as my first step in choosing happiness for myself. I chose to start telling myself stories that were very different from the stories I had when I arrived at Leader Dogs. I began telling myself about what was possible for my future, and stopped telling stories about what I once had, and had lost.

Around this time, I began to appreciate that facts are never as important as the stories we tell ourselves about those facts. Today, I believe that the single most important factor in determining your attitude toward a situation—and influencing the outcome—is *the story you choose to tell yourself about that situation*.

For example, some might say that blindness is a disability, and that's a fact. But that isn't the only fact about blindness, and it's far from the only story worth telling about blindness. If I want, I can choose to believe there's a purpose to my blindness. I'm blind because I'm strong enough to deal with it. I'm blind for a reason—to help others see their own potential.

I can go further than that. I can tell a story that I am grateful for the gift of blindness, for the perspective it's granted me, and for the challenges it's offered me, all of which have made me a better person. By reframing our deficits through the stories we tell ourselves, we can transform those deficits into our greatest assets. Choosing better stories about your weaknesses and frustrations can open the door to a self-fulfilling reality of strength, determination, and purpose.

But I wasn't there yet, not as a college student at the age of twenty-four. This was merely the time when the first seeds of these insights were sown.

ATLANTA BOUND

As a white man, I'd never experienced discrimination in my life, not really. Then came the first day of on-campus recruiting interviews at our business school. I'd made a series of appointments to sit down with recruiters at one of our big job fairs, but as I made my way from one interview to the next, I could hear the sounds of their dismay and disappointment when I approached. Set aside for a moment the laws prohibiting discrimination against the disabled. Some of these recruiters lacked manners and common human decency.

Right at the start of the interviews:

"Wait, oh my God, you can't see?"

"Uh, I didn't realize you have a dog."

"These jobs require travel."

"You know, this is probably not a good fit."

There was hesitation and stress in these voices before they'd completed a single word. I can read a room as well as any sighted person. Many of my interviewers I'd scheduled appointments with weren't interested in speaking with me.

I felt angry and hurt. It did not seem fair. But I couldn't let myself dwell on it. I told myself that if people don't get me, that's their loss. And if that's their company culture, I'm glad I won't be a part of it. That was the story I chose to tell myself. I hope the recruiters at these companies have better training and more open minds today than they did back then.

It was okay, though, because I didn't need all the jobs. I just needed one. And I ended up with two great offers from world-class companies. One was from Capital One, in its Tampa office, and another from the Atlanta offices of Accenture, which was then known as Andersen Consulting. To get those offers, I'd gone through many interviews, including a series of three- to four-hour exams to assess my analytical capabilities—and I'd evidently made the grade.

The Accenture offer came first, which gave me leverage when Capital One's offer arrived. I told Capital One that I had a limited time to respond to my other offer, and in twenty-four hours they upped my starting salary by almost 20 percent.

Atlanta is just three hours south of Knoxville, and I was somewhat familiar with it. But Tampa was a mystery. Over the winter holidays, my family and I went down to visit and found that we loved the area. We checked the routes to the Capital One offices, looked at some neighborhoods where Miles and I could go on our speed-walks, and came away very pleased at the prospect of living there.

The Capital One job would also be much easier to handle as a blind person. It was a regular office job with a daily routine and negligible travel requirements. At Accenture, however, I'd spend more than 75 percent of my time out of the office, consulting with clients. Most days, Miles and I would be finding our way through very unfamiliar surroundings. It would be more time-consuming and more stressful for the both of us.

The deciding factor, however, was in the job descriptions themselves. Capital One is a financial services company, and I preferred not to limit my future to that industry. Accenture would offer me a far wider variety of experiences to develop my skills. I'd be exposed to many different industries and all kinds of new technologies and management practices. Choosing Accenture over Capital One was certainly not the path of least resistance, but I had never been one to take the easier path just because it was easier. I knew even then that the right thing to do is seldom the easiest.

I accepted Accenture's offer in late December, and we agreed that I would start work at the beginning of February 2001. I discovered an apartment through a broker in the Sandy Springs area. It was a one-bedroom on the main floor of an apartment complex within walking distance of a Bally Total Fitness center and a bus stop that led right to MARTA, Atlanta's metro system.

During the month of January, I turned twenty-five. My friends rented a truck in Knoxville and helped me move in, and then my mother came down and helped me set up my kitchen, labeling the spices and boxes with bar codes that could be read with a scanner. She also helped label my clothes with tags shaped as squares, circles, and triangles for color coordination.

All during that month, I projected utter confidence when I was around my friends and family. But on the inside, I was teeming with fear. The thought of taking on a new city, a new job, a new profession, and having to travel daily on an unfamiliar transit system—all without being able to see—absolutely terrified me.

But I had other fears that loomed even larger: The fear of letting my parents down and the fear of letting myself down. We'd all done so much work to get me to this point that I felt compelled to march ahead. Like many other times before, my fear of not reaching my full potential drove me forward. I feared failing, but more than that, I feared not knowing what was possible unless I tried.

THREE

EXCUSES ARE FOR LOSERS

"GATE!"

"Gate!"

It was a Sunday morning, and Miles and I were standing on the ground level of the MARTA station at Dunwoody in the northern suburbs of Atlanta. I was teaching Miles to take me through the turnstile.

"Gate!" I said, slapping my hand on the turnstile's fare card reader. "Gate!"

Then I took a few steps back with Miles. I paused for a moment and then repeated the command. "Gate!"

Miles immediately walked to the turnstile and paused where I could find the card reader. I rewarded Miles with some exaggerated praise and lots of love—stroking his head and face (what dog handlers call "throwing a party"). I wanted Miles to remember what a good job he'd done, and where to take me whenever he heard me say "Gate!"

I had to throw Miles a lot of "parties" that first month in Atlanta. Back in Knoxville, Miles and I had been seen all over the campus for so long that he'd become a minor celebrity. He knew almost all my regular destinations—the lecture halls, the classrooms, the lunch spots, and the restrooms. He even knew his way around UT's gigantic football stadium, which is the fifth largest stadium in the world. Now we both had to learn our way around a new and much larger community.

During the week before I was scheduled to start at Accenture, I took Miles on several dry runs to Accenture's downtown offices on Peachtree Street. We practiced boarding the 87 bus on Roswell Road, riding for fifteen minutes, getting off at the Dunwoody MARTA station, passing through the turnstiles, taking the escalator up to the platform, and then boarding the Red Line going south to Peachtree Center. From there we navigated a maze of underground concourses and escalators on our way back up to the sidewalk level. Then it was just a few blocks to the Accenture building.

By the end of our second test run, both Miles and I knew the route cold. We'd leave our apartment at Plantation Creek on Roswell Road, and after crossing the street, I'd direct Miles to turn, and when I said, "Bus," Miles would take us to the bus stop. From there, I would direct Miles to the bus when it approached. But transit was still a new experience for both of us. Miles had to learn to find me a seat or let me know when there were no seats available. I also had to learn how to speak up for myself, growing more comfortable asking others for help. With time, I learned how to stand and "surf" with the rocking of the MARTA trains, or to brace myself by pressing one hand against the train car ceiling. I felt grateful to be tall.

Also during that week, I arranged with my Accenture recruiter for a guided tour of the floor where I'd be working. Learning the layout of a new office space takes me a little time. It helps to deliberately walk the premises and do some memory groundwork so that I can develop a 3D virtual reality image of the space inside my head. After that, I'm always aware at any moment of where I'm oriented, so I can move through the space, with Miles as my guide, as he's watching for obstacles such as other people. The walk-around also gave Miles the chance to learn pathways to the elevator bank, the cubicles, the kitchen area, and, of course, the men's room.

After eighteen months together at this point, Miles and I were a strong team. I could tell by his gait and the lack of hesitation in his movements when he had learned his way around the new place, which did not take long. Miles was a "show me once and that's enough" kind of dog. To this day, I've never encountered a smarter dog. He was extraordinary, even among the elite company of service dogs. He made Lassie look average.

My morning routine after waking up was to give Miles his food, refresh his water bowl, and then pull on some clothes and take him for his morning constitutional. Aside from his need to do his business, he also needed a healthy walk to stretch those long German shepherd legs before beginning his workday.

Back then, I worked out in the evenings instead of the early mornings. I'd return to the apartment with Miles from our morning walk, hop in the shower, and then get dressed for work. During my shower, I shaved my head, as I'd been doing every day since I was about twenty-two. I never could get my hair to behave, even when I was a kid. My hair was always straight and stiff, and it would never really cooperate, even with a military-style high-and-tight cut. Finally, I tried shaving my head one day, and I decided it was a good look for me. I was thankful I could pull off the look, because my hair never did me any favors. As it happened, my shaved head proved to be the most *practical* look for me, too. If you can't look in a mirror, the best way to make sure your hair isn't all over the place is to introduce it to a straight razor.

One of the many stories I told myself to help me come to terms with being blind was that I needed to figure out how to make blind look good. I'd always been a little vain about my appearance, so in my mind, the stakes were raised now that I was blind. Walking around an office in dark glasses with a large German shepherd meant that all eyes would be on us. It was more important than ever that we both look our best.

Accenture was a business-casual office, but I'm a firm believer in the wisdom of always dressing for the job you want, not the one you have. I made a habit of dressing "up," one level above the other analysts. I wanted to be stylish, yet practical. If I chose a look that was too coordinated, it would be more difficult to pull off the color coordination without eyesight. I had to strike a balance between style and simplicity.

I've always favored lightweight sunglasses with good eye coverage but not so large that they compromise style. I wore Oakley Halfwire Titaniums for years until they switched to a heavier alloy frame. Now I wear Ray-Bans with polarized Chromance lenses. I prefer polarized lenses because I still have some light perception and the glare from bright lights can be painful for me. If you've ever had your pupils dilated for an eye exam and then walked out into the sunlight, it's that kind of discomfort.

For client visits, I bought some well-designed suits, and had them fitted by a skilled tailor. I always sent them out to the cleaners after a single wearing. That way I could be sure they were never soiled, and that the pants creases were always freshly pressed. It was the same with all my other clothes. Everything goes out to the laundry after a single wearing and comes back cleaned and pressed. I had to be very disciplined about how and what I ate, taking care not to spill it on my clothing. If any piece of clothing gets a stain on it, I toss it.

It's ironic. I want to look good because I know from my own experience how we all judge people by the way they look. But now that I'm blind, I can also appreciate no longer needing to deal with such a useless and superficial distraction as other people's looks. When I hear someone speaking during a meeting today, I'm totally absorbed by the tone and content of that person's words. Other people around the table may notice that the speaker is having a bad hair day or wearing an ill-fitting suit, but from my perspective, everyone in the room could be wearing clown suits with full clown makeup. My blindness serves as a filter for what might be called irrelevant visual data. When it comes to using my active listening skills, being limited to only auditory input gives me an edge over everyone else.

I wanted blind to smell good, too, and at the time my favorite scent was Giorgio Armani cologne. Losing my vision really has enriched my experience of my other four senses. Without any visual information to process, my brain is better able to detect subtleties in sounds, smells, tastes, and textures. My sense of smell is so much sharper that I've got ten or fifteen different colognes to choose from, depending on my mood or the occasion. The unfortunate downside is that I also have a sharper nose for all kinds of unpleasant odors. If you skipped your shower this morning, I'll know.

THE WORKING WORLD

I began at Accenture as part of a crop of young beginner analysts, almost all of us fresh out of college, although there were a few who were making career changes in their late twenties and early thirties. A lot of the work Accenture did for clients involved writing computer code, so there were plenty of computer science majors in the analyst ranks. But coders weren't trained to understand business objectives and strategies in the code they were writing, so the teams assembled for a project usually had a mix of coders and people with business education backgrounds, like myself.

Accenture was a sink-or-swim environment. The hiring process picked us among the vast pool of applicants due to our brainpower and willingness to learn. None of us had much experience in what we were asked to do, and we were expected to learn on the fly. So, I did everything from website redesigns to marketing and financial modeling. The work is designed to help you grow by constantly putting you beyond your comfort zone. For me, outside the comfort zone was my everyday reality. Just shopping for food was outside my comfort zone. I constantly relied on trial and error, plus my innate stubborn hardheadedness, to keep overcoming obstacles, whether I was working on a financial model or trying to find ketchup at the deli at lunchtime.

Analyst cubicles were located in a kind of bullpen area, and because more than half of us were out visiting clients on any given day, there was no assigned seating. Each morning, I'd walk in and Miles had to find me an empty desk. Just as he had in Knoxville, Miles served as a great conversation icebreaker for get-to-know-you encounters, but I was determined that he not become a source of workplace distraction. There were a few dog-lovers in the office who had a hard time staying away from him, but those were also people most interested in learning about his unique training. In those conversations, I was able to subtly communicate that it was best for both me and Miles that he be regarded as just a fellow professional, and not as a cuddly office pet.

I never lost sight of the fact that I was a trailblazer, and that I was representing all blind people, not just myself. I was the first and only blind handler of a service dog that many would ever encounter in the workplace. I took my pioneer role seriously, to educate and set expectations so that the next blind person behind me would have an easier time than I did. It would be bad for me if I allowed Miles to become a distraction to my coworkers, but it would also be bad for the next blind job applicant if the hiring manager's impression of guide dogs was that they were unwanted nuisances.

My JAWS screen reader application worked fine with most of Accenture's software, but it went silent on one pretty critical piece: my weekly timesheets. Normally, as I navigate with the keyboard, JAWS tells me through my earbuds what my cursor is pointing at. But with the timesheets, where I had to log my hours, corresponding to each of my four or five assigned projects for each day of the week, JAWS couldn't tell me where the cursor was. I had to ask a colleague to help me devise a clumsy workaround, just to submit my time that week.

I was frustrated and a little embarrassed. I vowed to find a way to do this on my own, or my reputation in the office would suffer. I'd never learned to write computer code, but now was the time to try. I downloaded a book online, signed up for an email distribution list where folks shared ideas and suggestions, and was soon able to write some very rudimentary code that automated what I was doing for JAWS to read my timesheet information to me. It wasn't elegant, and it had some flaws, but it was good enough. More important, it taught me I could learn to write code if I ever had to. I just had to be ready for a lot of trial and error, a lot of slamming my shiny, shaved head against a wall until I broke through.

Work at Accenture was organized in project teams led by project managers, each of whom built each team the way players are chosen for a pickup basketball game. Over the months, some of the newer analysts distinguished themselves by getting picked by the top project managers for teams assigned to the highest-value, most prestigious client projects. I wasn't among those who distinguished themselves, and I couldn't be absolutely sure why. I knew my skills and experience were somewhat limited as a recent college graduate, and I wasn't highly trained or experienced in writing code. As a company, Accenture was very supportive, and even sent me to a training program for the technology, but over time I realized the project managers had reservations about how a blind guy would be able to keep up on their teams. As a result, I wasn't showing good numbers in my personal "billable utilization rate," the key indicator for assessing my ability to bring in revenue for the firm.

I just kept looking for an opportunity to shine. There was one project manager with whom I'd developed a rapport, and I talked to him about joining one of his projects for the testing phase of a new information technology product. Testing of this kind isn't a complex process, and I thought it would give me a chance to demonstrate my reliability. The project manager at first sounded very receptive to bringing me on. Then I heard nothing from him, so I stopped by his office to ask what was up.

Shane said: "I don't know what to tell you, Chad. There's a lot of new technology on this, and I'm not sure how it will work with your software."

He never said, "No," but he hemmed and hawed and I could tell his mind was made up. Weeks later, the project moved on without me and without a word from the project manager. It was such a huge letdown that I felt like I'd been kicked in the gut. It was a simple project, and he wasn't comfortable bringing me on board to do it. Was it workplace discrimination? Maybe, but it wasn't my style to go there. I refused to play that card. I preferred to focus my energy and effort on the things in my sphere of control.

When it came to public accommodations, though, I had zero hesitation. On St. Patrick's Day that year, after I'd only been in Atlanta for a few months, I made plans to meet friends at a down-town bar around 11:30 p.m. But when I got there, the bouncer guarding the door wouldn't let me in.

"Sorry, pal, no dogs." I could tell the bouncer was big because his voice was coming from above my head.

"I have friends in there," I said.

"It's too crowded," he said. "You don't want your dog in there."

"Come on, man," I said. "He's been in more bars than half the people in there."

"I told you, it's too crowded."

"I'm blind, and this is my guide dog," I said. "It's the law."

"Man, I cannot let you in. It's too crowded and busy in there," he said.

I could've tried to educate him about my rights under the Americans with Disabilities Act, which had been passed into law in 1990. Instead, I explained my position in terms he would be more likely to understand.

"Look," I said. "You do what you want, but if you don't let me in, I'm going to put you on the news."

"Sorry, man, I can't do it."

"Okay," I said. "Suit yourself," and I walked off.

So, the following Friday night, there I am with Miles on the Fox 5 Atlanta evening news, being interviewed in downtown Atlanta about the bar that illegally kept me from entering with my licensed and trained service dog.

About five years later, I ran into the bouncer in the Buckhead section of Atlanta. He remembered me well, because he'd been fired after the news report. All I could say was, "I tried to warn you, man."

MEETING EVIE

I had only lived in the Atlanta area for six months when I found a house to buy in Tucker, Georgia, about fifteen miles from downtown and very close to Stone Mountain, which is the most popular tourist attraction in the entire state. I was making pretty good money for a young, single guy, and I hated paying rent when I could be paying a mortgage.

It was a modest three-bedroom ranch house, and the commute was even more inconvenient than before, but it was affordable, a good investment, and it had a huge yard that allowed Miles to get outside and stretch his legs while getting some exercise. Eventually,

I got a couple of roommates to help me make my monthly payments. One was Josh, my best friend from high school who was looking to change housing arrangements in Atlanta, and the other was a trainer at the Bally Total Fitness center near my apartment who needed a place to live.

I liked having people around again, but the vibe in the house was much healthier than in my old South Knoxville group house. I chose my two roommates carefully because I wanted to live in an atmosphere that supported each of us pursuing his own goals and dreams. I've been a believer in what author Tim Ferriss says about friendships: "You are the average of the five people you associate with most, so do not underestimate the effects of your pessimistic, unambitious, or disorganized friends. If someone isn't making you stronger, they're making you weaker." Before you and I can become good friends, I want to make sure you won't be bringing down my average.

On Labor Day weekend, I made plans to meet up with Ken, another friend from high school who was visiting from out of town. Ken had recently gotten out of the Marines, and I was looking forward to taking him out for a night of carousing. We went to American Pie, a bar in the Sandy Springs area that was very popular with young singles.

At one point toward the end of the evening, Ken introduced me to a girl he'd been talking to, and then excused himself to get another drink. Her name was Evie, and she spoke with an accent that I couldn't quite place. She told me she was visiting from Brazil, where she grew up speaking Brazilian Portuguese, and had been a flight attendant for one of Brazil's airlines. She had moved to Atlanta temporarily for the sole purpose of improving her English so she could work on international routes. We only had about thirty minutes to talk before last call sounded from the bar. As we prepared to go our separate ways, I knew I wanted to see her again.

"So, you're gonna give me your number?" I asked.

"No," she said. "I don't give my number to strangers."

"Strangers?" I kidded her. "You know my name. I'm Chad. We're not strangers anymore."

"I don't think so." The tone of her voice was playful but firm.

Instead, she accepted my phone number, and then she was gone.

But there were a few things that impressed me about Evie. During the time we were talking, she never said a word about Miles, who was resting at my feet. No other young woman I'd met had ever done that before. Miles was always the easy icebreaking subject for women who knew they were the ones who needed to initiate contact if we were going to talk. Having women hitting on me all the time was fun, but the novelty had worn off a while ago. Increasingly, the conversations felt dull and predictable. Evie was the first young woman in a long time—maybe ever—to speak with me naturally and spontaneously about subjects other than Miles.

The other thing that impressed me about Evie was her courage. Her English wasn't perfect at the time, and yet, here she was, a single woman going it alone in an unfamiliar country, just to learn the language better. I thought that was a gutsy move. I felt like we could be kindred spirits, in a sense.

And yet, I'd given her the number to the landline at my house. (I hadn't given her my cell phone number because we'd only just met.) A week passed, and I still hadn't heard from her. I assumed that was that.

On Tuesday, September 11, I remember coming in a little late that morning and joining one of my project teams in our conference room. Just before 9:00 a.m., we heard a commotion outside the room and learned of the first airliner hitting the World Trade Center. We all tried logging onto CNN on our laptops, but the site kept crashing, so everyone scrambled to one of the TVs in the kitchen areas and watched the day's horrific events. An anguished cry went up in the room when the first tower fell. By then, we'd been told there were Accenture consultants working in that tower. We were all told to go home and didn't return until the following week.

I still remember the long, surreal transit ride home to Tucker that day, wondering when another aircraft might fall from the sky. I sat on my couch in shock over the next several days as news reports detailed the unforgettable horrors of the attacks. By the end of the week, five Accenture employees were counted among the 2,917 people who were dead or missing.

The layoffs at Accenture began before the month of September was over. The attacks on 9/11 had left the economy in such a state of shock that the market for consulting services evaporated. All our clients were trying to conserve cash, and when companies do that, high-priced technical consultants are usually among the first to go.

Every week brought a new wave of layoffs in the Accenture Atlanta offices, and I knew it was only a matter of time before my number was up. In eight months on the job, I felt like I'd never managed to overcome management's skepticism about my abilities. In a competitive environment, with all these young analysts selling themselves hard to get on the next hot project, the project managers seemed to look at me and wonder, "Can he do the job? What if he can't? Why take the risk?" My chargeable rate—the frequency clients were billed for my services—stayed low as a result, which I'm sure played a role in the decision to lay me off.

On the day I was finally let go, I had been spending most of my hours on a team developing a new product offering in the medical field. It was an internal project, meaning it wasn't bringing in any current revenue for Accenture. I was totally expendable.

Bam! They let me go. There I was. At home each day, a recent college graduate with bills to pay and approximately eight months of work experience in a hopelessly tough job market. I had about six weeks of severance pay, and about $10,000 in the bank because I was a habitual saver. I wasn't about to lose the house, but it was a tough time. I found myself ruminating about my frustrating experience at Accenture. Had I made the wrong choice? If I'd accepted Capital One's offer, I'd probably still have a job, because the financial services industry is a safer place during a recession. I'd be making more money, probably in a more supportive work environment, and living near the beaches in Tampa. I knew that was the wrong story to tell myself, but I struggled to find a better one in the short term.

Then Evie, the girl I met at American Pie over Labor Day weekend, left a friendly message on my voice mail. I'd thought of her from time to time, but it had been a few weeks since I'd given her my number, so I assumed I'd never hear from her again. I returned her call, a week passed, and then I got another message. We played telephone tag like this for a month before we finally reconnected.

We talked for hours. She was amazing, just as I had recalled: bold, passionate, energetic, articulate, and smart. She teased me about the night we met, saying she was surprised I could even remember our conversation because I'd had so much to drink. I assured her that I was only on my third beer, but there was a fun edge to her teasing that I really liked. We agreed to meet again for lunch and became friends. I was collecting unemployment by then, and Evie would occasionally drive me to job interviews.

It wasn't long before we started dating. I was aware at the time what a different experience it was to be dating without eyesight. When I could see, all my experiences had been filtered through the lens of my physical attraction. I would never get to know anyone who failed to make the cut of being attractive to me. The comedian Demetri Martin used to draw a chart onstage that showed how a woman's beauty was directly proportional to how many minutes he was willing to hear about her cat. I have to admit that I was like that.

Evie was and is a beautiful woman. But just as I'm undistracted by useless visual data in the workplace, I was never distracted by Evie's beauty the night I met her, when her intrepid and adventurous nature rang so clear for me in our conversation. I looked inside her that night in such a focused and intentional way only because I'm blind. To this day, I wonder if we would have had that same initial spark if I had been the guy I was when I could see.

Soon, our relationship became by far the most serious and intimate relationship I'd ever known. Evie was a kindred spirit, just as I'd felt when we met. I was falling in love, and it was both thrilling and frightening.

What would I do if this wonderful, charismatic woman really saw what I go through every day and decided that it wasn't for her? From her perspective, I wondered why would such a wonderful woman want to saddle herself with the burden of dealing with my baggage each day? Whenever we spent long stretches of time together, I noticed how much easier it was to move through the world as one half of a couple. That also scared me. What if I became too dependent on her, emotionally and otherwise, and then we broke up? All breakups are hard, but I began to worry that my daily survival skills were weakening in Evie's tender embraces. Would I ever get my edge back, my toughness and tenacity, if we broke up?

In this case, the story I told myself to overcome these fears was that I had to have faith—in myself, in the future, in Evie, and in us as a couple. I could see endless possibilities for our relationship, and I chose to think about them whenever my feelings of fear threatened to take over.

BRAZIL BOUND

There are few things as demoralizing as being unemployed. For me, the challenge was to avoid stewing over how Accenture's project managers had been biased against me and never gave me a chance to showcase my skills and talents. The unfairness of my time there hit me hard, but I also knew that I wasn't alone. A lot of people my age were out of work in the recession that hit after 9/11. Many of them were underemployed, making ends meet doing wage work.

I wasn't too proud to do physical labor, and I'd even worked in construction during college, but that was back when I could see. Most jobs that involve physical labor require eyesight. There are many good reasons that you've never seen a blind guy with a guide dog working on a construction site.

The unemployment rate for persons with disabilities is often above 70 percent, even in an ideal labor market. The best opportunities for workers with disabilities are when the economy is strong and qualified workers are scarce. Then the opposite happens when the economy weakens and there are more workers than there are available jobs. All of a sudden, hiring managers are flooded with résumés from overqualified job applicants. If you don't have the right degree, or a degree from the right university, or the right connections, it's likely your résumé is going to end up at the bottom of the pile.

Evie understood. Everyone understood. It didn't change the fact that I was in economic freefall, a December 2000 college graduate with almost no real job experience to differentiate me from the hordes of kids who graduated in May 2001. The only real differentiator was the fact that I went to job interviews with a German shepherd—and a lack of eyesight. I could feel myself being dragged toward the same swamp of self-pity that almost consumed me when I first lost my eyesight. No one would blame me if I folded and went back to Knoxville to regroup. Whatever excuses I had to offer, the world would accept, simply because most cannot imagine what it is like to be blind. Everyone would give me a pass if I asked for it.

My thoughts ran back to my cousin Mark's dose of tough love in that driveway in South Knoxville. He warned me what would happen to me if I lowered my expectations for myself and hung out with people who did the same. I would become a loser. Sure, the Accenture layoff could suggest that I wasn't cut out for this fast-paced, high-tech environment. But if I let those challenges lower my expectations for myself, then I'd be the only one losing. For the first time, I came to understand that excuses are for losers—and I was *not* a loser.

Excuses. Are. For. Losers. All excuses. Of all kinds. Yes, I was caught in a seemingly hopeless situation beyond my control. There were no jobs, and I had a mortgage to pay. Everything I'd accomplished in my first year after graduation was about to crash and burn. But by then I'd already overcome enough troubles to know that this situation—like any situation—wouldn't get solved by making excuses. Joblessness was putting everything I had at risk, but I'd only gotten this far by not making any room for excuses. Now was not the time to start.

It's a lesson I'd learned over and over in the three years since I'd gone blind. Nothing blocking my way ever got shoved aside by an excuse. So what if my current problem was a big one beyond my control? I'd been pushed to my limits many times before. Each time, I had to plow ahead and resist the urge to make excuses. Sometimes that worked. More often, it didn't. So, I'd have to try again. And again. And again.

We can't always control what happens to us. We're not responsible every time a bolt from the blue puts us flat on our backs. But we're all accountable for how we respond. This simple insight is one of the greatest gifts my blindness has granted me. Over the years, my ability to remedy any bad situation has always depended upon my sheer unwillingness to make excuses.

At this point in my life, I was still testing this idea, but I already knew I couldn't sit around and bathe in the comfort of Evie's sympathy. Playing that card would have gotten old for both of us pretty fast, so I kept doing what I'd been doing since I returned home from Leader Dogs training. I did the hard work of earning respect instead of expecting pity from others, which has always proven to be the healthier choice anyway—for me and everyone around me.

My job hunt dragged on for months, and when the holidays rolled around, I had the humbling experience of returning home unemployed. I'd told my mother in advance that if she wanted to give me something for Christmas, I could really use a new laptop computer to help me with my job search. I gave the exact model and specifications: a Dell Inspiron outfitted with the latest CPU, and the maximum amount of RAM that Dell could fit in the chassis. My mother shipped the laptop to me in Atlanta before Christmas had even arrived, just so I could get on with my job search. It was a workhorse of a computer that I would rely on for the next ten years.

In February 2002, I finally landed a job with the Georgia Department of Labor. I would be a disability benefits case officer, which was ironic on a number of levels. I was a disabled worker processing the paperwork for people who claimed to be too disabled to work. The vast majority of them, of course, were truly too disabled to work. But it was also part of my job to weed out the occasional slackers who claimed to have a series of mysterious maladies preventing them from rejoining the workplace. That's right, the excuses-are-for-losers guy spent a portion of every day looking over disability claims, trying to determine which were genuine and which were just excuses from losers.

How did I land a job when jobs were so scarce? I suspect the credit goes to Accenture. There is one thing every hiring manager knows about someone with Accenture on his résumé. That person has made it through Accenture's rigorous interviewing process. Hire an Accenture alum, and you know you've got someone with the aptitude and attitude needed to learn quickly on the job and do things right. For months, I'd bemoaned my time at Accenture as a wasted opportunity. The new job gave me a story to tell about how that time had served me well, after all.

My time at the Department of Labor told me that I didn't want to stay there very long. Although the people were amazing, the pace of work was glacial and the antiquated information technology systems weren't helping matters. I had felt kind of run over by

Accenture's fast-paced, highly competitive work culture, but the dragging days at the Department of Labor reminded me that I belonged in an Accenture-type environment—I just had to sharpen my coding and technical expertise, so I could effectively compete in the game. I wanted to get back to a place where the work I did was creative and innovative, and where continuous improvement and learning were highly valued. There was very little of that at the Department of Labor.

One day, Miles got violently ill at the office and threw up all over the place. Immediately, I left work and hustled him off to the vet. I was alarmed because Miles *never* got sick. The vet determined that someone had given Miles people food, probably as a treat when I was focused on work. By the time Miles and I got back in the cab, the vet visit had cost me $500.

Everybody on our floor knew how furious I was about the incident. When I found scattered leftovers of human food by his bed in the office, my blood just about boiled over. All service dogs are fed a restricted diet that is very stable for their digestive systems. They also get only a measured amount of food each day to prevent them from getting overweight. I had explained to everyone that part of Miles's training was to be desensitized to the smells of human food so that he would never be distracted while doing his work. I guessed that one or more of the dog lovers in the office had assumed they knew better—and then they learned otherwise.

The next day, one of the physicians who worked for the department pulled me aside and handed me an envelope with cash in it. "This is from Rita," he said, referring to one of the women on our floor. "She feels terrible about what happened to Miles and took up a little collection." I never tried to find out the truth, but

I always assumed it was Rita who had snuck Miles the little treats that nearly killed him. She probably felt too guilty to give me the envelope herself.

While waiting for Miles to be treated in the veterinarian's office, I was reminded how deeply dependent I was on him. What if Miles had to stay overnight to recuperate? Obviously, I would have to sack out for the night in the waiting room. How else would I get home or even find my front door? If your car needs to be in the shop for a few days, the dealer can give you a loaner car. But there are no guide dogs on loan at the vet.

It wasn't long after I landed my new job that Evie told me she'd booked a flight home to Brazil in March. I'd always known this day would come. When we'd first met six months earlier, she told me all about her plans to return to Brazil in the spring. But I still hated the idea of us being apart.

"After you've been with your family for a month or so, I'll come visit," I told her.

"You will love Brazil," she said. "But I have to warn you. It's going to be much harder for you there."

"I can deal with it," I said. "You know me." From my perspective, nothing worth doing had ever come easy to me, so why would Brazil be different?

"Yes, but you don't really get it," Evie said. "No one wants dogs in their restaurants and hotels in Brazil. And there are no special laws for blind people like there are here." She really wanted me to love Brazil, and she was afraid I'd hate the restrictions on our time there just because I'm blind.

"It sounds like a knife fight in a phone booth," I said. It was a phrase I used from time to time to describe a difficult experience. I gave her a big smile.

"That's what it will be," she said. "As long as you know in advance."

I got it. Evie didn't want me to be blindsided. But I'm blindsided every day, all the time. I'd much rather be blindsided with her in Brazil than alone at my desk at the Department of Labor any day.

I booked my flight for May and took Miles to get all his vaccinations and paperwork for international travel. All through the weeks leading up to the trip, she and I phoned back and forth. We decided we wanted to travel a lot while I was there, and many of her updates contained news of her finding yet another hotel or restaurant that agreed to accept Miles as a guest. Piece by piece, Evie spent weeks creating a special itinerary for us—a guide dog's guide to travel in Brazil.

I made one special preparation for the trip that I didn't tell Evie about. I bought an engagement ring. When I met Evie's parents, I planned to ask Evie's father for her hand in marriage.

The ring cost much more than the salary of any low-level Georgia civil servant could afford, even though I'd gotten it through a wholesaler friend of my brother's. When I told my dad about it, he warned me, "Son, you're setting the bar awfully high for the next purchase." Wise words indeed, but it was a once-in-a-lifetime purchase, and I was never one to do anything halfway.

Evie's parents spoke no English, so I planned to speak in very practiced Portuguese while requesting her father's permission to marry her. An occasion like that deserves a ring that won't disappoint.

Evie and I had often spoken of making future plans together, but we had never discussed if we would marry. There was one night I'll never forget, when we were chatting about something in my master bathroom, and Evie blurted out, "Chad, I want to have your babies!" It scared me in that moment. At twenty-six, I didn't feel at all ready for the responsibilities of raising a family. But it was a solid tip-off that Evie was ready to say "yes" if I were to pop the question.

Evie's younger sister, Zenith, and brother, Marcelo, had both moved to Georgia and were living in her apartment while she was in Brazil. I asked Zenith if she'd meet with me because I needed her advice.

"Do you think Evie will like this?" I asked, as I removed the diamond ring from its velvet pouch.

I could hear Zenith gasp.

"Oh, my," she said. "Of course, Chad. She'll love it! It's beautiful."

I listened for hesitation in her voice and heard none. Sisters talk, and if Evie's sister thought I might be jumping the gun with a marriage proposal, I would have heard it in her tone of voice. Instead, her cool demeanor assured me this was the right thing to do.

As my travel day drew closer—as did all the knife fights in telephone booths that awaited me in Brazil—I recognized that I had no end of understandable excuses for not taking this trip. But excuses are for losers, and I was no loser. I was the luckiest man alive. I was going to Brazil to see my bride-to-be and to humbly ask her parents to bless the most important commitment I would ever make in my life.

FOUR

LIFE BEGINS OUTSIDE YOUR COMFORT ZONE

"PURPOSE OF YOUR VISIT?"

I was in the line for *Nada a declarar* (Nothing to declare) at São Paulo's airport.

"Visiting my fiancée and her family," I said, feeling glad the customs official spoke English.

"You have nothing to declare?"

"Nothing to declare," I lied.

Miles stood quietly next to me as the customs official stamped my passport and told me to move along. Tucked into a secret compartment on the inside of Miles's harness was Evie's engagement ring.

You may wonder why I turned my guide dog into a jewel mule that day.

Sadly, I'd heard nothing but bad things about Brazilian customs. A college professor of mine who had once lived in Brazil told me that, on one trip, it had taken him more than a week to figure out who to bribe and for how much before he could get his belongings out of the airport. Evie also told me how notoriously corrupt Brazil can be, how the officials there might ask you for a *cafezinho*—a small coffee—and you're supposed to know they're asking for a bribe.

So I decided weeks before leaving that I didn't want any part of that. I never even bothered to find out what the duty charges would be. I was not going to risk declaring it, because at best I'd have to bribe them. At worst, they'd take advantage and rob me blind-er.

I'd used Miles in the past to smuggle a flask into football games, hiding it in a pocket on the back where the harness crosses his spine. I laughed at the thought of anyone daring to get close enough to Miles and his teeth, just to pat down his harness. Even TSA airport security was either too deferential to me or too scared of Miles to search his harness. So, that's where I hid Evie's ring.

After sitting on the runway at Hartsfield-Jackson Atlanta International Airport for three hours, I'd taken a ten-hour overnight flight from Atlanta to São Paulo, and tossed and turned in the seat the whole time. It had been months since I'd seen Evie and I was anxious about our reunion. I was also nervous about meeting her mother and father for the first time. Then there was the ring. If I lost it or if it were confiscated, its cost would be the least of my concerns. The hardest loss to swallow would be the lost opportunity to propose to Evie. And what about that? What if she said no?

I was feeling vulnerable in ways that I had never known. The final uncertainty was how I'd be received in Brazil. I'd learned that Brazilian families tend to be ashamed of disabled family members, some even keeping them out of view. If you ever see a blind man out in public in Brazil, it's not uncommon to find him panhandling on the corner. German shepherds are plentiful in Brazil, but usually as military, K-9, or guard dogs. What would the Brazilians make of me and Miles, with Evie on my arm? In every way imaginable, I felt far, far outside of my comfort zone.

I took a short nap during my layover in São Paulo, and as I began to wake, I could hear the faint sound of music in the background. Groggy from almost no sleep, I was stretched out on a couch and could hear "Old Blue Eyes" singing, and for a second, I wondered where I was. Frank Sinatra was belting "New York, New York," and I paused for a second because, for a moment, I wondered where I was. *Wait, I am in Brazil, right?* I thought. I'd expected to hear the sounds of samba or bossa nova when we landed, not Sinatra.

After my four-hour layover in São Paulo, I boarded a flight to Goiânia, where Evie's friends and family lived in the Brazilian interior, not far from the national capital, Brasilia. Evie and her parents were waiting for me at the airport, and there was an emotional scene as we all met and embraced.

I was obviously wrung out from the twenty-hour trip, and Evie and her parents were eager to help me with my luggage. But there was a carry-on bag I wouldn't part with, which puzzled Evie. I couldn't tell her why I insisted on carrying it myself: I'd transferred her engagement ring from Miles's harness to an outside compartment on the bag, plus it had the paperwork for the ring just in case

I was asked about it. I didn't want Evie to accidentally stumble across her big surprise.

Our plans were to spend the next five days at the hot springs resort town of Caldas Novas, home to one of the largest geothermal aquifers on earth. Evie had taken great care and effort to find accommodations that would welcome her American boyfriend and his German shepherd. It hadn't been easy. In some cases, it had taken weeks for Evie to convince restaurateurs and resort managers to make this one special exception to their ban on dogs on their properties.

At the start of our third day there, I found the perfect moment to ask Evie's father for his permission to wed Evie. We were eating breakfast at a restaurant and Evie excused herself to go to the buffet, leaving me alone with her parents.

I turned to Evie's father and said, *"Posso me casar com sua filha, senhor?"* May I marry your daughter, sir?

Evie's father burst out laughing with joy.

"Claro!" he exclaimed. *"Claro, meu filho!"* Of course, my son. It was truly music to my ears.

Both of Evie's parents were laughing, and I could hear them sniffing back their tears of joy. Evie returned from the buffet and said, "Why are you all crying?"

"We're just happy," her father said in Portuguese.

"Yes," her mother said. "Very happy."

Evie was quiet, the way she gets when she's suspicious something's up. I could tell she wasn't satisfied with our answers, but she let it go.

I put off popping the question until our last day at Caldas Novas, when we were bathing in one of the hot springs. I hid the ring in Miles's harness before we headed for the springs that evening. Once we were soaking in the warm, frothing waters,

I called Miles over, and presented the ring to Evie. She was so shocked she went totally silent, which left me in suspense since I couldn't read her facial expression. Not until she fell into my arms could I be sure of her answer.

The following day, we all traveled to Rio Verde, where Evie's parents lived. The four-day visit became a festive engagement party for Evie and me. The family's roots were in Italy, and her grandmother was an amazing cook. I have a special weakness for Italian food, and this was some of the best I'd ever tasted. We awoke every morning to the smell of baking bread, and as a real aficionado of fine cuisine, it was heaven on earth.

The rest of our time in Brazil was spent in Rio de Janeiro and Ilha Grande ("Big Island"), a scenic, unspoiled island near Rio that was kept in its natural state for centuries because it was first a leper colony and then a high-security prison—the Alcatraz of Brazil. When we were in Rio, we stayed at the world-famous Copacabana Beach, where signs in our hotel warned guests not to wear jewelry on the beach—day or night.

I was well aware of Rio's serious crime problem, which made me uneasy to think about walking the streets there with Evie. What if someone tried to grab Evie's bag or threatened us? How could I protect my bride-to-be? One thing I hadn't considered was the effect Miles would have on passersby. To people in Brazil, I didn't register as a vulnerable blind guy with his guide dog. Instead, I looked like a tall, ripped cop from the States, wearing dark shades and walking his trained K-9 attack dog.

Everyone steered clear of us when they saw us coming, and Evie thoroughly enjoyed playing it up. When a stranger asked if our dog bites, she would smile and say, *"Somente quando ele fecha a boca."* Only when he closes his mouth.

Of course, taxis, restaurants, and hotels also steered clear of us. Miles was blocked everywhere we went. We ended up outside a Brazilian steakhouse for almost an hour trying to get in, wrangling with one host, and then another, Evie speaking in rapid-fire Portuguese that I could barely understand.

Finally, a nice older gentleman behind us interceded and identified himself as a police officer. "You need to let these people in," he said.

The restaurant's host responded that he was worried about having a dog near the food.

"Don't worry," the man said with reassuring, authoritative tone. "It's a trained dog. Trust me. I'm a police officer."

The host relented, and after we were seated, we thanked the officer for speaking up on our behalf. Evie said, "It's so good to have the support of the police in situations like these."

"Police?" the man replied. "Oh, I'm not the police. I just thought what they were doing was wrong. I felt bad for you two."

HELLO VIRGINIA

Evie and I returned to Atlanta from Brazil on the same flight that week, and over the summer of 2002, we started spending so much time at each other's places that she eventually moved in with me at my house in Tucker. I would ask her to pack "just in case clothes" for when she wanted to stay over, and the next thing we knew, almost all her clothes were at my house.

At this point, I was anxious to work in tech again and get back to writing code, so I picked up my job hunt. In the past year, I had taught myself a lot about coding because I was a highly motivated learner—I *had* to learn code to get JAWS to work with all the programs and applications I wanted to use.

Before the summer was over, I got what I felt at the time was a perfect job offer. The Bartimaeus Group specialized in making workplace software accessible, particularly with agencies and companies that needed to comply with government regulations for accessibility. If there was a drawback, it was that the job was in the Washington, D.C., area.

Evie and I talked about it, and we both thought it made sense for Evie to move with me, but this would mean she would have to quit her job and lose her health benefits. So, we fixed the benefits problem by getting married. We went over to the courthouse in Decatur and got legally hitched just a few days before we made the move to the D.C. area. We planned to get married anyway, so this just made it legal a little sooner.

It wasn't until the following May that we held our wedding ceremony, on a beautiful Tennessee hillside in Gatlinburg, with about a hundred friends and family. Evie's mother and father were unable to attend because they were unsuccessful in convincing officials to approve their visa. However, her sister and brother, along with her friends, came in from Atlanta to attend. That evening after the reception, Evie told me that the wife of one of my friends had approached her at the reception and asked her how she really felt about marrying me. I'm sure she was asking what many others there were thinking.

"It might not be easy," she said to Evie, "with him being blind, you know?"

"I love him," Evie replied. "We'll work it out."

Fall 2002 was a nerve-wracking time to live in the D.C. area, because that October was when the infamous Beltway sniper attacks riveted the nation and launched a three-week regionwide police manhunt that left everyone walking on eggshells. One day, when Evie and I were leaving the Home Depot in Falls Church, Virginia, I clowned around for Evie's amusement, darting right and left on my way out the front door. "I'm making it a little harder for them to shoot me," I joked. A few days later, our blood ran cold as we read that sniper fire had killed a forty-seven-year-old woman in that same parking lot, mere yards from where we'd been walking. An eerie feeling of angst overcame us both as we pondered what could've happened had our fortunes been slightly different. It felt like we narrowly escaped death. By the time the two snipers were finally caught on October 24, they'd killed ten people and critically injured three others in the previous twenty-two days.

It seemed like the most recent move—we'd been through so many at this point—might have been the toughest on Miles. I could not get him to slow down, even though I'd tried every technique there was. Miles always wanted to run, which forced me to hold him back with both hands, with one on the leash and the other on the harness. This posed a problem, because I couldn't carry a briefcase or an overnight bag with both hands occupied.

I continued to wonder if there was a way to slow Miles down, so I asked Leader Dogs to send a trainer to visit for an evaluation. It was around Christmastime, and we went to the very crowded local mall in McLean, Virginia, during the heart of holiday shopping season to give Miles a real test. We could all see that Miles was working flawlessly. It was as if he knew he was being judged. Miles had a real problem with walking too fast, but his mind was sharp. He nimbly stopped, started, paused, dipped, and dodged our way through an absolute maze of humanity that was stomping through the mall—we never even bumped into anyone, which is more than most sighted people could say under those circumstances. He worked flawlessly—it was utter perfection.

When I recognized Miles was on to our checkup, I told the trainer, "Let me show you what Miles can do when he wants to." I then simply addressed Miles and said, "Miles. Find the car." Miles turned and retraced our steps perfectly, down one wing of the mall, then another corridor, then through the department store where we entered, past several entrance doors, out the same doors we'd come through, across the parking lot, into the parking garage, down a flight of stairs, across two rows of parked cars, in the parking garage, and down a long row of cars, stopping at the liftgate of our SUV. The entire time, all I had said was, "Find the car." We couldn't even remember where we'd parked, but Miles knew. He was extraordinary, even among service dogs. The trainer said he'd never seen anything like it.

"Chad, I don't know what to tell you," he said. "We can't give you a better dog than this. If I were you, I'd just hold on a little tighter." So, that's what I did for the next few years with Miles—I held on a little tighter. It was a good thing that I worked out every day, because I needed that strength to handle Miles and his tendency to sprint.

We had rented a one-bedroom apartment, in Falls Church, just about fifteen minutes from the Bartimaeus Group offices. However, that didn't mean much to me because I spent most of my days working on-site at client offices all over the region. We had a driver who shuttled me and Miles to and from many of my job locations, but I also grew accustomed to standing out in the bitter cold waiting on buses and trains.

In working with clients, I found my blindness gave me a big advantage because I had to really know the systems and standards in order to write software for myself. It was easier for me to do a thorough job, because I was reliant on the same software myself.

In the spring of 2003, I traveled with the company's team to Los Angeles to attend the annual assistive technology conference sponsored by California State University at Northridge. The conference's exhibitor hall was the largest of its kind, attracting companies from all over the world, and it raised my awareness of just how fast assistive technologies were evolving. It really was a great time to be blind, and it seemed to be getting better by the day.

There was a buzz in the exhibit hall one morning that the legendary singer-songwriter Stevie Wonder and his team were there to sample some of the new products on display. I asked Yara, one of my colleagues from Bartimaeus, to take me where Stevie was because I wanted to talk to him. When we got near, Stevie was surrounded by fans asking for pictures and autographs, but I didn't want to come across as a needy fan. Instead, I thought about how I could help him.

Yara leaned in and introduced us, and I remember the feel of his soft handshake. I told him briefly about our company and asked if there were applications in his music studio that he'd want to use more effectively. He told me that, yes, there was a software package he'd been having trouble with. Would I be interested in looking at it? I could hardly believe my ears. I told him I'd be glad to and gave him my card. Later on, Steve Jones, Stevie's computer engineer, tracked me down (I'm easy to spot in most crowds), and we talked for half an hour about what Stevie needed in his studio.

I ended up extending my stay in town, just so I could visit Wonderland, Stevie Wonder's legendary Hollywood studio. Stevie came by to pick me up at my hotel in an SUV driven by Stevie's bodyguard. After lunch at Roscoe's House of Chicken and Waffles, we stopped by a music store to pick up some specialized software Steve and I had discussed. At every stop we made, Stevie was approached by adoring fans, and he was always gracious and welcoming. Even when we were eating at Roscoe's, he'd put down his fork, chat with his fans, and pose for photos. He never so much as grumbled under his breath about the interruptions.

On the way to Wonderland, Stevie and I chatted about my eye condition, among other things. He promised to put me in touch with a surgeon who might be able to restore some of my vision. We were interrupted by a phone call. It was Ray Charles. The two of them talked about possibly getting together that evening, and for a moment I began to hope that, maybe, just maybe, I'd be lucky enough to hang out with two living legends on the same day. But Ray told Stevie he wasn't feeling well enough to join us at the studio, so, unfortunately, destiny had other plans.

The rest of the day stretched long into the night. I later learned that Stevie enjoys being in his studio so much, he spends days there without leaving. Steve, the engineer, and I went over the studio setup and working with music software made me feel like a kid in a candy store. Music is huge in my life: I love everything from Kenny Chesney to 2Pac, from Prince to Adele, from Dave Matthews Band to Jack Johnson, and of course, from Stevie Wonder to Ray Charles. Chances are, if I'm not in a business meeting then I'm listening to music—even when I'm wearing my JAWS earpiece. Music keeps me in flow.

It took us about an hour to figure out what I needed to do with the software when I got back to the office in Virginia. Then, for the rest of the evening, we hung out. We watched a basketball game, and then switched to the news, where it happened to be the first night of the bombing of Baghdad in the 2003 Iraq War.

"Shock and Awe" was front and center for us in prime time on the television. We were all in disbelief that our nation was at war. Dinner in the form of Chinese food arrived. Stevie was noodling on the piano while I discussed martial arts with his bodyguard. The memory of the evening seems like a dream now. I finally had to leave at 3:00 a.m., because I had an early flight. I ended up not even sleeping that night. I went from the studio to my hotel, took a quick shower, grabbed my bags, and headed to the airport to hop on my flight back to the East Coast.

In the following weeks, nothing ever came of my work with Wonderland because all my Windows-based accessibility software was incompatible with the Mac platform preferred by the recording industry. I spoke with Stevie a few more times on the phone, but never did get the name of his surgeon.

That turned out to be the highlight of my time at Bartimaeus. Soon, I realized that my future was best aligned with the business world, and not so deeply engrained in the technical world. Although I was learning a lot, I began to wonder if spending so much time in this niche market would pigeonhole me and close me off from the business mainstream. It was my third job since leaving college, and I still didn't know what it would be like to stay in one place for a while.

By August 2003, I'd made my next stop at Bender Consulting, where I was hired as an account manager in charge of a team of twenty-five generally young consultants working on-site at a huge information technology company called CSC (Computer Sciences Corporation). The most interesting work I did there was as a market intelligence specialist, researching industries and companies that would need CSC's commercial outsourcing services, and mining information for the CSC executive sales teams.

Sometimes I'd contribute market research to help prepare competitive bids for gigantic outsourcing deals for CSC's technology services. I was very impressed with the CSC pricing teams that worked on these bids. They were really brilliant people, some of the smartest I've ever met.

After about two years, I began to think about jumping ship and asking CSC if I could join one of their pricing teams. As much as I loved working at Bender, the job had become a real grind. I was putting in ninety hours a week serving three different roles as manager of twenty people, accessibility guru, and market intelligence analyst. Bender would have been a great place to learn and grow if I had wanted a management career, but my love of numbers suggested to me that a field related to financial analysis was where I really belonged.

I had to pitch hard to join the CSC pricing team. When I first asked the people there about a job, they were hesitant, but I didn't take it personally. "I know it's a stretch to take a chance, you don't know if I can do the job," I said, "so kick the tires, give me a test, throw some financial models at me, ask me some questions." Because I was in their offices every day as a consultant for Bender, they already knew me and my work ethic and reputation. Eventually, they put me through some tests, just as I asked, and then they offered me a job as a pricing analyst.

Ash, my new boss at CSC, took me under his wing and really played an active role in coaching me about all the dynamics of pricing a bid. I'd never had a mentor or a sponsor in my previous jobs, and I often thought that I might have fared better at Accenture if I had. After Ash, I knew I'd never want to be in any workplace without one.

Many companies fail to win large technology bids, I learned, because they assume their cost estimate forms the foundation for their bid price. In fact, the bid price must be based on whatever the open market currently pays for the service. You have to research that market price, then adjust your company's costs to fit the market. Otherwise, your company will never win a bid.

This is true for selling anything, really. You may think your house is worth $300,000 because of all the improvements you've put into it, but if all the comparable houses in your neighborhood are selling for $200,000, you will never get your price. All your improvements are only worth what the market says they're worth, news that can come as a shock if you were expecting more.

"Shock" would sometimes describe the reaction of CSC engineers when I'd inform them that others were doing roughly the same work for just 80 or 90 percent of their proposed cost. It meant they would have to figure out a way to do more with less, or we wouldn't win the bid. At the same time, the salespeople kept pushing for lower and lower bid prices. You need to balance having a competitive price versus winning a large deal that is priced too aggressively. Salespeople usually just want to win the bid, even if winning the bid would obligate the company to bleed red ink for years to come.

On the pricing team, I was told that we've done our job when both the salespeople and the engineering teams were equally angry at us. Our job was to push both of them outside their comfort zones and navigate the ensuing tension. Tension is healthy. Without it we can never be sure we're submitting the best possible product.

The most interesting thing about the job was that no one really knows the answers for sure. Once you're done triangulating the masses of data points, arriving at specific numbers for the bid is as much art as science. You're only dealing in probabilities, so the final decisions are still made on gut instinct about the numbers you feel good about.

Our financial modeling system for pricing these mega deals was very efficient, but unusually complex. Each bid was managed by a bid team of about fifty associates, who included people from lines of business, accounts, and pricing teams. All of us could work concurrently on 115 separate spreadsheet files that were all linked to one another. Each file had its own set of worksheets and calculations. There were a hundred cost models with another fifteen rollup files that we used to instantly summarize these multibillion-dollar deals. And, on a global deal, there may be more than six hundred models to manage different file-sets for different worldwide geographies.

When I joined the pricing team, I was told the setup is so complex that it usually takes people about a year to comfortably lead pricing for a deal. I respected the sophistication required, but I was not aiming for usual. I made it my goal to beat the twelve-month benchmark, and I ended up leading my first deal after only eight months on the job.

It was sink or swim for me to get there. I spent nights and weekends writing more than ten thousand lines of software code so JAWS would work more efficiently and effectively with Microsoft Excel. I managed to tie Excel to JAWS so I could programmatically access information about the spreadsheet in a way that also allowed me to automate certain tasks that our pricing team had always done manually.

One macro code I wrote allowed a pricing person to complete what had been a two-hour task in less than five minutes. Under deadline pressure, it saved our team significant amounts of time that could be better spent analyzing the deal or negotiating with our deal partners.

My career advancement depended upon achieving this kind of mastery of Microsoft Excel. Master carpenters will tell you that, if they have to consciously think about how to use their tools, then they have not yet mastered their craft. True mastery is arriving at a point where the tool fades out of consciousness. A master carpenter envisions the desired outcome, and the strokes required to get there just happen naturally and effortlessly. That's the kind of relationship I was developing with Excel. I spent many long hours creating a sophisticated web of hundreds of interlinked Excel files, each with dozens of worksheets, thousands of calculations, and reliable results—all with the end in mind and, of course, in complete darkness.

Envisaging and building these spreadsheet architectures was easier because of—not despite—my blindness. I get a lot of practice in abstract thinking because everyday life for me is an exercise in abstract conceptualization of my surroundings. When I work with complex, abstract ideas, I can form mental images without needing to shift my attention away from concrete visual perception. And no doubt my experience with engineering software gave me a clear advantage when building these financial models. If you can build a complex piece of software, building a financial model comes pretty easily.

My coding skills stayed sharp because, after I left Bartimaeus, I'd never stopped writing accessibility code. Every time I saw a new application that interested me, like Windows Messenger, I'd write the code needed to make it operate through JAWS. I also made some additional cash on the side by taking on consulting work, customizing JAWS accessibility for clients with very specific software needs, much as I'd investigated for Stevie Wonder.

I admit there were times when I wasn't much fun to be around. Quiet weekday evenings at home would find Evie reading or watching TV, and I'd be sitting with my laptop, writing and rewriting computer code with JAWS yammering in my earpiece. I was teaching myself coding, getting better at it all the time, the way some people master video games. Every job was different, and there was no guidebook for what I was doing. When I'd find myself unable to get the code to run as intended, I'd recall how my high school wrestling coach taught us persistence: "Just grab it and growl."

"What are you *doing*?" Evie would sometimes ask me, because I'd disappear into a world of my own for hours on end, furiously typing code instructions. Occasionally, when a good-sized consulting check would arrive in the mail, I'd say triumphantly, "See? That's what I've been doing."

For one client, I found a way to make JAWS function much more accurately and efficiently with Lotus Notes, a popular business application owned by IBM at the time. After several iterations, I was able to sell licenses for the software. We put that money into our first investment property, a vacation condo in Florida.

I was with CSC for almost two years, and I might have stayed there much longer but for one nagging concern: I was working for CSC's *commercial* outsourcing team in Northern Virginia, and my team did no business with the federal government, which is the largest buyer of IT products and services in the world. I was building a track record in a valuable area of expertise, developing my own distinctive way to do pricing research and financial modeling. But I knew that I could never really unlock my full potential without any experience in the specialized field of federal bidding and contracting. I decided the time had come to make a move, but I would wait until the right opportunity came along.

FAREWELL TO MILES

Sometime around 2006, we noticed that Miles started shaking while he was working in harness. I could feel it, and Evie and others could see it. Miles liked to work, so he'd get excited when it was time to hop in the car and go. But then when we'd arrive at some destination, like a restaurant, Miles couldn't calm down. He would lie at my feet and start shaking nervously and uncontrollably.

This was the first time ever that Miles was a source of concern. He was always so steady—a complete rock.

We kept an eye on the behavior for a few months, and there was no improvement. It was as though Miles still had his head and his heart in the work, but his nerves were shot. I had worked Miles hard for more than seven years, and I guessed that all the travel and house relocations had taken a toll on him. Miles was eight at the time, approaching retirement age for guide dogs, and I decided it was time for him to enjoy his golden years.

Some blind people who retire their guide dogs keep them as pets, but for Miles that was never an option. Miles had severe separation anxiety, and with both Evie and me out of the house at work all day, it would have been intolerable for him. Plus, we could never leave him alone unless we wanted to come home to a wrecked house. It would be torture for Miles and us both. Instead, I used my network to find a retired couple in nearby Fauquier County who had plenty of time to give Miles all the love and attention he deserved. Their names were Bob and Nancy, and they lived on a farm, allowing Miles all the room he needed to run to his heart's delight.

Our final weekend with Miles was a tearful one. It was like saying goodbye to a family member who was moving far away. We decided that with Miles now entering civilian life, we'd send him off with some of the delicious people food he'd been missing out on. Evie prepared Miles a sixteen-ounce rib-eye steak, Brazilian style, cooked over an open flame with rock salt. It was the dead of winter, and nobody hates the cold more than Evie, but she stood there out on our patio working the grill in subfreezing weather. "I'm only doing this for him," she said. "Don't think this is for you. This is not going to be a regular thing."

At first, Miles was wary of the special meal we'd prepared for him. Despite my encouragement, he had to overcome a lifetime of training to avoid people food. Maybe he thought it was a setup? He stared at the bowl for about ten minutes, then finally dug in and eagerly devoured it all.

The next day, Bob and Nancy came to pick up Miles and carry him off to their farm, where he'd be surrounded by fields, able to live out the rest of his life chasing rabbits and deer. Bob and Nancy even found a job for Miles, after I'd told them how he felt a need to have a purpose. A local school had Miles come by every week as a teaching aid for children with reading disabilities. These children, who were so self-conscious about reading aloud in class, loved sprawling out on the floor and reading to Miles. He always wore a special harness when he was with them, so he knew he was working. He enjoyed being the center of attention, and I like to think it gave him the same feeling of service that he enjoyed as a guide, the sense of purpose we all need and crave.

That's my final memory of Miles. I didn't keep in touch with Bob and Nancy much longer after that, because it would have broken my heart over the years to hear about his eventual decline and passing. Miles was my rescue dog. He rescued me from one of the darkest times of my life and every day afterward he literally led me toward the light. He took me from the rock-bottom of my life all the way to the boardroom. For seven years he gave me all his devotion and loyalty until he finally had no more to give. In my mind's eye, Miles is forever listening with rapt attention to the soft voices of children reading to him or racing with mad abandon through the green rolling farmland of Northern Virginia.

THE BIG FEDERAL BID

In October 2007, I got the job offer I'd been hoping for, working for a federal contractor on a big bid. A large technology company hired me to join a team tasked with winning a competitive bid for the company's largest ongoing contract—all the information technology services for TSA, the federal Transportation Security Administration.

This was the challenge I wanted. It was a multibillion-dollar contract, with service operations in dozens of locations all over the country. My role as pricing manager was to benchmark the going prices for services of the kind that TSA required, and then influence the bid team to use my recommended pricing targets and business strategy.

It took me less than thirty days at this company to realize I'd gotten off the train at the wrong stop. My research was showing that many of the assumptions they used for pricing their services were making the company noncompetitive. I had some creative ideas for making adjustments so these prices wouldn't sink the big TSA contract, but many of them fell on deaf ears. I don't want to say I was usually the smartest guy in the room, but I was among the savviest when it came to pricing and pricing strategy. It was hard to find someone I felt I could learn from while there, and that's not what I wanted. I wanted to be around people who would challenge my assumptions and help me raise my game. I also wanted to *win.* I'd learned that while pricing a bid is a very complex process, the end result couldn't be simpler. You either win or you lose. They were playing this game like they didn't want to win, or just didn't know how. I felt like telling everyone that they could save a lot of time and money if they didn't bother bidding at all.

When corporate recruiters called, I explained to a few of them that I wasn't happy at my current employer, but that I felt it would be a betrayal to leave before I'd done the work I was hired for. I had a particularly good conversation with Ben, an executive from a company called SRA. Ben was on the next level. He enjoyed the game involved with preparing a bid, which you have to if you want to win, especially big ones with the federal government. SRA was one of the best at doing that, and I knew I could learn a lot by working there.

Ben respected my reasons for staying with the company. He even put a positive spin on it, as a learning opportunity for me. "Look, there's no deal in the market right now that's bigger or more important than the one that you're working on," he said. "There's no better place for you to be. Stay there, do what you need to do, learn what you can. And when the time comes, we'll reengage."

My new guide dog during this time was a German shepherd named Romeo. He was even bigger than Miles and had a little attitude problem that bubbled up from time to time. He'd occasionally growl at strangers, something Miles had never done, not even once. When Romeo did it, I'd have to give him a firm leash correction immediately, which sometimes startled onlookers. Occasionally, I'd hear an animal lover tell me, "You shouldn't do that to your dog."

I had to suppress my impulse to tell these busybodies in very colorful language to mind their own business. Instead, I remembered my role as a goodwill ambassador for all guide dog handlers.

"What if he growled like that at your child?" I'd ask, in as diplomatic a tone as I could muster. "Would you say the same thing?" I thanked them for their concern and assured them I knew exactly what kind of behavior correction my dog required. Romeo and I had both been taught by some of the best dog trainers on the planet.

To say I missed Miles would be an understatement. After Miles retired to his new home, I had a black Labrador retriever named Blue for a while. Believe it or not, my chief motive in changing breeds was vanity. German shepherds are nicknamed *German Shedders*, and I wanted to avoid picking unsightly tan hair from the dark suits I wore to the office. A black Lab seemed like a smart solution, but Blue lasted less than six months. He was clever, but so playful that he wanted to play more than he wanted to work. One day, passing by a sports store at a local mall, Blue saw a kid bouncing a basketball and couldn't help himself. He just yanked me toward the ball and jumped on top of it. "Boom! I got it!" his body language said as he uncontrollably seized the ball.

That was it for Blue. I vowed at that point to go back to what I knew—German shepherds. They are worth the trouble, even with all the hair they shed. As soon as I could, I took Blue back to Leader Dogs, where I was matched with Romeo. I don't know how Romeo got the edge to his personality—perhaps it was in his genetic makeup—because I never could train it out of him. Even at Leader Dogs, there were several occasions when he just pounced on other dogs. It was something I never could have imagined Miles doing even once. I wondered if it had been someone's little joke to call him Romeo, because this Romeo was certainly no lover. He was a fighter.

By August 2008, I'd been at my new employer for about ten months, and we had just submitted our preliminary RFI (request for information). I did my job but felt nothing but doom about the project. The company numbers I was working with were so far out of whack with the marketplace reality my research was turning up, there didn't seem to be any hope of us ever winning this contract. But there was nothing I could do about that, and the occasional phone calls from Ben at SRA helped keep my spirits up. Ben's advice to me was good advice for anyone in this kind of difficult work situation: learn what you can, and do what you can to benefit your self-development, while putting your best foot forward for the company.

One afternoon that August, I headed off for lunch at the nearby Champs Sports Bar with my usual lunch buddies. When we returned, the atmosphere was so subdued I could feel that something had changed. People were speaking in hushed tones and the phones were weirdly quiet. It felt like a funeral, and sure enough, it was our project that had died. While we were at lunch, word had come down from the government that we had been disqualified from bidding on the TSA contract—before we had even submitted our price. This came as a total shock because no one saw it coming. For seven years, TSA had been the biggest government account on our books. No one could imagine that we wouldn't even get a chance to rebid on its renewal.

I called Ben that afternoon, very discreetly. We arranged to meet at SRA's offices the following week. We'd never met in person before, and I wondered if Romeo would growl at him and sabotage what felt like the most important job interview of my life.

Ben explained to me during the interview that it wouldn't be enough for me to simply present my numbers—I'd have to justify them, especially if I was recommending a strategy and efficiencies in how work gets done, which I often did. Pitching is all I ever do, I told him. I did it successfully at CSC. Where I was, I'd gotten plenty of practice trying to explain the same thing a dozen different ways, and eventually getting some traction with folks, but to most my approach was entirely foreign. That made him laugh.

When I opened up my laptop, it felt good to walk Ben through a few of the more recent financial models I'd developed. Some of them enabled us to quickly test for and execute cost reallocations between services that would give us a bidding advantage, allowing us to strategically include dollars in certain service areas based on projected demand. Ben had a lot of questions, which was great. They were the right questions, the kinds of questions that no one was currently asking me.

That's when Ben stopped and asked me if I was really sure I could do the job. I assured him that the job was so easy that "I could do it with my eyes closed." His laughter suggested to me that I'd finally get the job I'd been exploring for nearly a year, and that my transition was coming soon. I felt excited about the opportunity to find a home at SRA, where I could lead, learn, and contribute in a meaningful way to SRA, our financials, and create career opportunities for our associates by winning large deals.

FAMILY MATTERS

On the day I went to meet with Ben, Evie was about five months pregnant, and we were expecting a little girl to join our lives by Christmas. We had been talking about having children ever since we began dating in 2001, but we'd always kept putting off the date to begin trying. Finances were our main consideration. As I moved from one job to another, and my salary kept climbing, we bought a townhouse in 2003, and then flipped it in 2006 for a newly built home, where we'd begin our family. It was a spacious, four-bedroom home in Kirkpatrick Farms, a new development in Loudon County, Virginia. The developer of Kirkpatrick Farms had given the streets colorful names like Mindful Court and Laughter Drive. Our house was on Experience Way.

Evie had always wanted children, but when I was younger, I wasn't so sure. I liked our life as a couple, and I'd always felt children were too much work and trouble. I'd hear parents struggle to control their kids when we were out shopping or eating in a restaurant, and I tried to imagine what a hassle that would be.

Besides, I liked our life as a couple. It was great to be married and just have each other without any other obligations. We loved going to concerts during those years. When Dave Matthews came to play in Northern Virginia, we'd go with friends and make a day of it, tailgating in the parking lot, barbecuing for hours before the concert even began. We went to a Jack Johnson concert during these years, eagerly expecting to hear "our song"—the soft, sweet ballad "Better Together." As the concert neared its end, we grew concerned that Jack was going to skip it. I figured that from the front-row disabled section, it was easy for me to let Jack know how we felt. As his second encore was drawing to a close, I stood up, and with every ounce of strength I bellowed, "BETTER TOGETHER!" To this day, we still laugh about how Jack Johnson played our song as his unplanned final encore, and how I hadn't been escorted from the arena for disrupting the show.

Maybe no one is ever truly ready for children, but at age thirty-three, I felt as ready as I'd ever be. We had the house, and I'd landed the new job at a nice salary bump. I figured this would be the time. The only other consideration was that I would be certain to pass on the recessive gene for retinitis pigmentosa to our offspring, and if Evie happened to carry the same recessive gene, there was a chance our children would get my condition. We knew of no genetic testing at the time to allay our fears, but one thing we had going for us was that Evie was descended from Italian and Iberian stock, far from my own Anglo-Irish ancestry, making it extremely unlikely that Evie had the same gene mutation that both my parents had. Evie herself was fatalistic about the question. She used to say that whatever was in the cards for us, we would deal with it, and we would be great at it.

In the early months of 2008, Evie had a checkup scheduled with her gynecologist, and she suggested having the doctor remove her IUD. That's how we decided to begin trying to start a family. It didn't take very long. In the middle of the night in March, she returned from a late-night visit to the bathroom and woke me up with the exciting news that she was pregnant. She hadn't even told me she'd gotten a home pregnancy kit (it's very easy to keep secrets from a blind spouse), so I wasn't sure if she was kidding or not. She had to put the little plastic test dish in my hand to convince me.

We were so elated, we had a hard time going back to sleep. It was a far-off, someday moment we'd been anticipating in our dreams, and we couldn't believe it was right in front of us. We were awash with joy, excitement, concern, and intense anxiety. It was a huge responsibility. Was I ready to be a dad?

In the following months, during one of our visits to the obstetrician's office, we heard our daughter's heartbeat for the very first time, and we almost melted. It was a feeling of joy like none I'd ever known before. I made a recording of the heartbeat and played it back countless times. The sound of our daughter's pulse of life overjoyed my heart.

As the delivery date neared, Evie's mother came from Brazil to stay with us for a few months, so she could be there for Evie. One night, when Evie was still six weeks away from her delivery date, she told me she felt like she might be going into labor. Something just didn't feel right. We went to Arlington Hospital, about forty minutes away in light traffic, and after Evie was admitted, they decided to put her on a medication to delay her labor until full term. They wanted to give our daughter a few more weeks to develop. They sent us home in the early morning hours of that Wednesday in early November.

When we awoke later that morning to get ready for work, Evie felt like something was still not right. There was genuine concern in her voice, and it troubled me because of the fear in her tone. When we were about ready to leave, Evie said she had the distinct feeling she was going into labor again. She had an appointment with her obstetrician, so she called him to ask for advice. He told her to skip the appointment, and he'd meet us at the Arlington

Hospital. Everyone felt an intense sense of panic, because we knew we'd have to fight morning rush hour, and it might take us ninety minutes or more to get there.

I had to stay calm. If Evie saw me looking scared and panicky, it could send us both into a spiral at the very moment when we had to keep our heads. I calmly loaded Romeo into the back of our Jeep Grand Cherokee, and we all hopped in the vehicle to head into the notoriously insane Washington, D.C., workday morning traffic.

Evie showed incredible courage and determination as we started and stopped, again and again, on the backed-up Dulles Toll Road. I could hear Evie sucking wind, holding it together behind the wheel.

I grew increasingly impatient, and then I got angry. I was angry about the situation. Why hadn't they told us last night this might happen? We should never have come home. But most of all, I was angry that I could do nothing to help, that Evie was forced to drive herself to the hospital because she has a blind husband.

After about twenty minutes, I decided I couldn't just sit there with nothing to do.

I called 911.

Operator: "This is 911. What is your emergency?"

Me: "Hey, my wife is in labor, and we're eastbound on the Dulles Toll Road. Traffic is at a standstill, and she needs medical attention, now!"

Operator: "Is she okay? Does she feel like her life is in danger?"

Me: "She is okay, but she needs a doctor, now."

Operator: "Okay, sir, calm down."

Me: "I cannot calm down. She is driving our car to the hospital in Arlington."

Operator: "Please, sir, have her pull over to the side of the interstate, and we'll send help."

Me: "Evie, they want you to pull over, and they'll send help."

Evie: "Pull over? *No way!* Are they sending a helicopter or an ambulance?"

Me: "Ma'am, are you sending a helicopter to help us?"

Operator: "We're sending an ambulance, sir."

Me: "Evie, they want us to pull over and wait for an ambulance."

Evie: "No way, not in this traffic. They'll never get to us."

Me: "Ma'am, we cannot do that. Please tell the ambulance to come look for us."

Operator: "How will they find you?"

Me: "We're in a silver Jeep Grand Cherokee, eastbound on the Dulles Toll Road, driving in the breakdown lane."

Operator: "Sir! Please pull over and wait for the ambulance."

Me: "We can't wait. We won't make it."

I'd never felt so helpless as I did when I hung up. The ambulance wasn't coming. I tried to imagine what I would do if Evie's water broke in the car, when we were stalled in traffic. Would I try to deliver our baby with the help of a phone assist? Would I jump out with Romeo and try to flag someone down to help us? It was a shattering feeling, to know that the moment might call for a hero, and I wasn't sure if I could step up to the task. I was of no use in this situation. All I could do was be there for Evie, feel the stop-and-start lurching of our Jeep, hear the sounds of Evie gasping with pain and blaring the horn.

Finally, we made it off the interstate. Arlington was just a few miles away, but each stoplight now felt like an eternity. Finally, I heard Evie say, "Okay, I see the hospital."

For the first time all morning, I could begin to relax.

FIVE

WHO WANTS IT MORE? NOBODY!

"SHE'S SO BEAUTIFUL AND SO TINY, CHAD."

"She's strong," I said. "At least, based on the sound of her screams."

I was at Evie's bedside, holding her hand after the delivery. At 2:30 p.m. on November 12, 2008, Juliana Marie Foster had entered this world six weeks ahead of schedule and weighing only four pounds.

It had been a difficult day, and for a few minutes, things were touch-and-go. Evie had been admitted as soon as we arrived at the hospital, and both she and the baby were hooked up to have their vital signs closely monitored. We were told right off that today was the day Juliana would be born, which worried us because the pregnancy was only at week thirty-four. A baby's lungs are not considered fully developed until week thirty-six, and lung treatments for premature babies have been known to cause lifelong disabilities.

At a little past 2:00 p.m., we were all sitting quietly in Evie's room—just Evie, her mother, and me. Then the monitors detected a sudden drop in the fetal heart rate. Instantly, a swarm of doctors, nurses, and assistants converged on Evie and swept her off to the ER for a caesarean section. I trailed behind, holding Evie's mother by my side. No more than fifteen minutes later, Juliana was delivered.

I will never forget the sweet relief I felt when I heard Juliana emerge from the womb, screaming at the top of her obviously high-functioning lungs! That was 2008, and she's rarely stopped talking ever since.

Only later did the doctors discover the reason for the early delivery. Evie's placenta had signs of calcium buildup, which constrains the flow of blood through the umbilical cord. Since the fetus relies on that blood flow for all its nutrition, Juliana had actually begun to lose weight in the womb. By using medications to delay the delivery, we were effectively starving her.

But Juliana busted out of there early anyway—just so she could get a good meal.

That's *my* daughter!

STRATEGIC THINKING

Juliana's birth left me overcome with emotion. It was as if someone had taken my heart out of my body and placed it gently in this tiny angelic being. It was terrifying to feel so exposed and vulnerable. A tidal wave of emotion washed over me when she laid there in the hospital. Its weight nearly took me off of my feet. I thought back to all the times I'd brushed off my parents when they said, "You don't know how worried we were about you." Now I finally understood the depths of their love and concern for their visually impaired son. For the first time, I realized all the worry I must have caused them every time I walked out the door.

For the next day or so Juliana was confined to an incubator, and she spent a total of seventeen days at Arlington Hospital's newborn intensive care unit (NICU, or "nick-you," as everyone called it). Each morning, Evie and her mom would go to the NICU with my parents, who also gave me rides back and forth to work. I'd arrive at the NICU later in the evening, although I often had to take work calls and even fire up my laptop at the hospital just so I could answer questions from my team at SRA.

I had just started at SRA, and I was working on a large bid with the Department of the Treasury. For one of the first times ever, SRA was bidding on a managed-services contract, and I was the most qualified pricing person to do the job, because of my prior experience with managed services since learning at CSC. As we put together our bid, I had to teach the entire deal team about the complexities of managed services, how to capture hardware cost, software cost, how to model labor cost and account for capital outlay, depreciation, and cash flow projections along with financial hurdle rates such as return on investment, internal rate of return, return on invested capital, cash flow breakeven point, and operating margin, all while laying out a strategy for winning the work *and* remaining profitable. I created the modeling architecture used by our entire bid team, the pricing targets, cost profile, and underlying business assumptions and risk profile for our proposal that totaled into the hundreds of millions of dollars.

That might seem like a lot for anyone to tackle, and it's especially challenging when that person has only been in the company for a few weeks—dealing with the practical matters of relationship building required to succeed, not to mention the unspoken ideas as to what this blind guy could actually do. A few of my new friends in the office quietly advised me to be careful, that there were people at SRA expecting me to fail. I wasn't given any names, but once I had my head on a swivel, I could tell who might be having trouble dealing with me. I got it. Everyone is fighting for a slice of the pie in corporate America. There's only so much pie to go around, and some people find it difficult to digest losing their slice to a blind guy.

I had been at SRA for several weeks when I first learned that, prior to my start date, the entire marketing and sales organization had been put through sensitivity training to prepare for my arrival.

One day, I was speaking with Yvonne, one of my coworkers, when she blurted out, "I know I'm not supposed to ask, but why are you wearing a watch, Chad?"

"Wait," I said. "What do you mean, you're not supposed to ask?" Then Yvonne told me about the trainer who'd been brought in to advise everyone how to speak with me, and to avoid asking me questions that might hurt my feelings.

I know the company held this training with only the best of intentions, but I could not stop laughing as Yvonne filled me in with one story after another from the sessions.

I've learned with time that I have no way of controlling what people think when they see me. Several times over the years, I've been on a business trip, eating alone in a restaurant with my guide dog at my feet, and the server advises me that some kind patron in the restaurant paid my bill. The first few times this happened, I felt somewhat insulted, thinking that someone had assumed I was a charity case deserving of pity. Then I worried that maybe there was something wrong with my clothes that day, and I looked down and out. With time, I realized, I have no idea what stories people project onto me. Maybe they assume I'm an injured veteran. Or they like my dog. Or maybe seeing a blind man suddenly makes them count their blessings, and they pay my bill out of a sense of gratitude.

Who knows? I finally decided that all I can do about other people's perceptions is to tell myself a better story about them. These well-intended souls were paying it forward and who was I to stop them? So, I've taken their cue and done the same many times over. SRA's leadership went to extra effort to consider my feelings with sensitivity training, which was extremely thoughtful but entirely unnecessary. They had no way of knowing they were getting the Chad E. Foster brand of blind. Whenever I sense there's some tension in the room about my condition, I'm the first to drop a blind joke to help put everyone at ease. I've found that the best way to deal with the proverbial elephant in the room is to invite everyone to have a good laugh about it.

Back when I could see, I remember all too well how I assumed that my blindness would limit my horizons. I remember feeling scared that I might not be able to support myself, much less ever excel at anything. So, how can I fault any of my coworkers for their skepticism? The work at SRA was very complex and involved searching online through thousands of documents for nuggets of useful competitive intelligence. We needed to research the customer, the contract, our competitors, congressional appropriations, and all kinds of other project funding artifacts. We were looking for needles in haystacks, sometimes just a few words or an interesting number buried in three thousand pages of data. It's natural to doubt how I could possibly do any job that was so challenging for people with functioning vision.

What the doubters failed to grasp was how I was using my disadvantage to my advantage. My blindness had forced me to develop a deep understanding of technology, which I used to create intelligent search algorithms that would comb through thousands of pages in just a few minutes. While my sighted peers would spend a week or more visually skimming the same volume of material, possibly missing what my algorithms turned up, I would fire off my code, walk down the hall to get a cup of coffee from our break room, and return minutes later to a summary report of the findings.

The purpose of this research was to help us build our recommendations for strategic pricing. We had to examine each government program's previous bids, pull them apart, and reverse engineer them so that they would reveal the award tendencies of the procurement teams as well as our competitors. It helped us understand what the customer valued and prioritized. The process is similar to SAT test preparation: you'll score higher if you understand how you're being graded. Whoever the customers are, it's important to know how they tend to make their choices on the basis of price, and then set your prices accordingly.

For example, most gas stations with convenience stores set their gasoline prices very competitively, because a low price attracts customers. That's a strategic pricing choice, to earn reduced profit margins on gasoline sales while selling coffee and snacks at higher margins. In a similar way, our bid had a greater chance of success if we put a very competitive price on services that our research revealed were most likely to influence the decision makers. Other services that our customer didn't regard as quite so critical to assessing our bid were areas where we could make a more acceptable rate of return.

It all sounds very reasonable, but it's actually quite difficult to execute. The pricing recommendations we produced often called for some of our operating units to meet high performance standards at low levels of staffing. I'd get a lot of questions and sometimes very intense pushback at meetings where our recommendations were aired. Some executives on the team responsible for executing the project (if we won the bid) would tell me that they wanted to vomit when they saw what was expected of them.

Once, when we were bidding on a software project for the US military, I gave a presentation that pointed out how one particular area of operations would require significantly leaner staffing if we wanted to win the contract. Tom, SRA's chief growth officer, asked the head of that department what he thought.

"Hold on a second," the executive replied. "I need to get the gun out of my mouth."

I felt the people in the room fidget in their seats uneasily. Some forced a laugh, but everyone knew this wasn't funny. He continued: "There is no way we can be expected to execute this plan. We can't reduce staffing by that much and still deliver."

I explained that we had worked with this particular contracts office before, so our direct experience told us the low costs were necessary. "I'm sorry it's a tight one," I said, "but according to the way these guys award bids, this is the way to win the work."

When my research reveals that the market rate for one of our services is lower than what we normally charge, it often inspires some creative thinking about how to get our costs down. This wasn't one of those times. We didn't follow my guidelines on this bid, and we didn't win the bid.

But we won plenty of other work, including the big Justice Department software bid, which was a huge triumph for the company, and for me personally. This time, we celebrated the win at Ruth's Chris Steak House. For this win party, I'd arrived late, and drink tickets were in short supply by the time I got there.

I asked the host of the party, an executive named Steve, if he could find me a drink ticket so I could get some scotch on the rocks. Steve announced aloud to the room, "No way we can let the brains behind our winning pricing strategy go thirsty!" He returned minutes later and presented me with an enormous glass of premium scotch.

Steve was very happy for the new deal we'd added to his portfolio, and he knew I was instrumental to our pricing strategy and how I had negotiated with many of our bid team partners to get everyone's cost in the right place. We ended up having several more drinks that evening, and I don't want to venture a guess how costly our bill was, but at some point we upgraded from premium scotch to Johnny Walker Blue Label. For the record, SRA did not pick up the final tab—that was a bill split among friends after many months of long workdays well into the evenings.

ATLANTA

We brought home little Juliana seventeen days after she was born, when she was finally plump enough to leave her incubator. To hold her and hear her tiny beating heart gave me an indescribable feeling of joy. I wanted to spend a few weeks at home with her, but I was still new at SRA and had not yet accumulated any time off. So, I approached Ben with a proposal to use a company policy where I could borrow some paid time off from a few colleagues, who'd told me they'd be happy to help out.

"I think we can figure this out among ourselves," Ben said. He'd seen my hard work and dedication, putting in long hours on my first deal with the company. "Look, here's what I want you to do. For the next two, three weeks, whatever you want, I want you to sit at home and do some 'strategic thinking' on some of these deals we have coming up. Check your email once a day or every other day, maybe take a call from the office now and then. Let's just handle it like that."

I learned an important leadership lesson from Ben that day. Whenever you're going through a difficult time in life, or a life-changing event—like a birth or a death in the family—you will never forget the people who stepped up and helped you carry the load. If you can be that kind of leader to the people who report to you, they will run through walls for you after that. After the three weeks I spent at home, thanks to Ben, there was nothing I wouldn't do for him for the next eight years we worked together. Ben set the standard for the kind of leader I aspire to be.

I wanted to help care for our new baby, which entailed a whole new series of learning experiences. Squirmy babies are hard to bathe with your eyes open, so imagine trying it with your eyes closed. Changing diapers wasn't easy or fun—a dirty job that sometimes required a return trip to the bathtub.

Careful organization is the secret to doing anything without looking. Everything is based on system, process, and touch. First, establish the landmarks with your hands—in this case, the bathtub, the water, and the baby. Then locate a place for everything, and be careful to put everything back in the exact same spot where you found it. Soap. Shampoo. Towel. Fresh diaper. Clothes. Everything needs a permanent spot.

Life can be difficult for blind people when they live with those who are not, because sighted people rarely feel the need to put things back where they belong. I wanted to help feed Juliana on her bottle, but the formula never seemed to be where I'd left it. Then, once we started feeding her baby food, I learned the hard way that it wasn't worth the trouble. I couldn't locate Juliana's mouth with the spoon, and she was too tiny and uncoordinated to reach for the spoon with her mouth. Things got messy pretty quick, so I mainly left feeding to my wife and family until Juliana was a little older.

My parents returned to Tennessee soon after we brought Juliana home from the hospital, but Evie's mom stayed for a few more months, which was very helpful. Evie drove me to and from work each day, and when she did, Evie's mom stayed home with Juliana. Things changed for us when Evie's mom left for Atlanta to visit Evie's sister and brother. That's when our new everyday routine began. Every morning, Evie had to gather together all Juliana's baby gear for the thirty-minute ride to work, and then strap Juliana into her baby seat in the middle of the second row of seats in our Jeep Grand Cherokee, with Romeo riding in the cargo area.

At the time, I was handling cradle-to-grave pricing on each deal—strategy, pricing target, actual pricing of the bid, and submission files—so it was hard to work more than one deal at a time. But sometimes I'd have to start a new deal before the other one was ready, and then some deals would come back from the customer for revisions. All of a sudden, I could go from one deal to four, which made for many long nights and weekends of work.

Things were not always easy between Evie and me. We argued often. We both have very strong personalities, and a new baby in the house presented all kinds of new opportunities for clashing. The daily routine was very stressful, and Evie was lonely, with few friends in Northern Virginia, and a husband who rarely had downtime to share.

One day, I was all alone on a conference call in my office when my cell phone dinged.

I pulled the phone near to hear a cold electronic voice read me a text from Evie.

"I have left the house with Juliana. Please only talk to me through my attorney."

Evie had left me and taken the baby. My body went numb. I was in shock.

I hung up the conference call on the spot and texted Lu, one of my best friends at work. Together we went in to see Ben, who was incredibly supportive. He arranged to get me an expense account for travel, so I could take a car service between home and work. Several days later, Ben told me I'd be welcome to stay with him and his family at their house in Arlington if it would make my life easier. How many corporate executives would be willing to make that kind of offer to an employee *and* his dog? I'm sure he cleared it with Beth, his wife, and their three children, which says a lot about the entire Gieseman family.

What followed were some lonely nights in our big empty house. "Why is this happening to me?" That question I used to ask myself as a victim now challenged me to tell a better story, one that would serve me. How can I explain what's happened in a positive way?

I knew that Evie wanted to move back to Atlanta to be close to her family. She needed her family around, and her sister was her best friend in the world. She never liked Northern Virginia to begin with, and being cooped up with a baby all day was making her stir-crazy.

I always promised her we'd move back to Atlanta someday, but the truth was I didn't really see how we could. All the work at SRA revolved around our main customer, the federal government. My expertise would be wasted if I took a different job in Atlanta. Besides, I liked working at SRA. I valued the mentorship I was getting from Ben and several others in the company. To me, SRA felt like family.

I spent many long hours ruminating over what I should do. I couldn't bear to throw in the towel on our marriage, not before I knew I'd tried everything to resolve the situation. The one question that kept coming back to me was this: What would I say to Juliana thirty years from now? Would I be able to tell her honestly that I tried everything I could to keep our family together? Until I could answer that question, I knew I needed to do more.

Within a day or two, I concluded that my only choice was to ask Ben for a transfer to SRA's Atlanta location. I hated the idea of leaving the Northern Virginia headquarters, but I saw no other way to save my marriage and reunite my family. I knew that I was putting my career at risk. The upward trajectory I was on owed a lot to the key mentoring relationships I had with Ben, Joe, and several other senior executives. Also, informal face-to-face conversations seemed a key part of my success so far. Persuading our operations teams to lean into some of these difficult pricing strategies was not a simple transactional process. It required tact, person-to-person assurances, and some knowledge of office politics. My reputation for hard work and winning contracts was solid, but a lot of my influence was tied to the relationships I'd developed with my deal teams. I knew I might be derailing my career, but it was a choice I had to make.

Ben didn't bat an eye when I asked for the transfer. He got it approved the next day, and within a week he'd made all the arrangements for setting up a private office for me at SRA's Atlanta location. My mind was blown by his support and generosity. With my personal life in free fall, Ben made sure my work life had a solid foundation.

My friend Lu helped us pack up and move to suburban Atlanta, where I'd rented a small apartment. For the next several weeks, Evie and I communicated only when we swapped time with Juliana. Evie wanted us to go to marital counseling before she'd even consider reconciling. Over the years, she had occasionally dropped hints about marital counseling, but I'd always shot it down, judging it as unnecessary. Now it was the only way we could get back together. For the following five months, we spent one hour per week with a marriage counselor who, like Evie, was a Brazilian-born woman. If I wanted us to reunite as a family, Evie required home field advantage throughout the process.

The fact is that all cross-cultural marriages have some difficulties, and marriages with one disabled spouse also have specific kinds of problems. We needed help with both. From the day we met, I'd admired Evie's strong personality, and it hadn't really occurred to me in our eight years together how our two strong personalities would clash, especially once we had the added stress of raising a child. Now it was time for me to change my tune or suffer the consequences. I had to humble myself, and learn to listen in ways I hadn't before, even though I'd always considered myself an excellent listener.

Those were long, lonely months—living alone again, away from my friends in D.C., and getting accustomed to working remotely from Atlanta. My daily meditation became more important to me than ever. I had started using mindfulness years earlier, at first just for a few minutes each morning after my workout, to bring some calm and clarity to my mind before work. During this time alone in Atlanta, my commitment to it grew deeper, and now I practice mindfulness a few times each day, usually in ten-minute intervals. I use an app called Headspace, which is like having a monk in your pocket.

I also worked out harder than I had in years, with rock, rap, and house music pounding in my earbuds. Two kinds of positive self-talk got me through it all. The first was something I learned from working out. I'd make a ritual of asking myself *Who wants it more?* My answer was always the same. *Nobody!* I would get my family back together. I would make this new, challenging work situation work for me. All because I was telling myself that nobody wants it more than me.

The second line of self-talk is one that's always been a kind of mantra of excellence for me. *If it were easy, everyone could do it.* Outstanding performance always relies on accepting that it's bound to be more difficult than other people are willing to tolerate. "The last three or four reps is what makes the muscle grow," said Arnold Schwarzenegger, in one of my favorite quotes from his bodybuilding days. He said, "This area of pain divides the champion from someone else who is not a champion. That's what most people lack, having the guts to go on and just say they'll go through the pain no matter what happens."

Sometimes, the only difference between success and failure is picking yourself up off the floor, just one more time. I was knocked down and got back up a lot during the summer of 2010. After three months in an apartment that didn't suit me or Romeo very well,

I bought a house in the beautiful Atlanta suburb of Dunwoody in August. By October, Evie and Juliana joined me. We were a family again, with a new understanding of how we were going to live and grow together.

HEADHUNTED

Working from Atlanta turned out better than I could have imagined. My reputation at headquarters was solid enough that I didn't need nearly as much face-to-face meeting time as I had assumed. My work spoke for itself. When needed, I could hop a flight to headquarters in the early morning and be back by late night, though I usually stayed for the week so I could catch up with the folks I worked with on a regular basis.

I also traveled to customer sites to negotiate and discuss large contracts. One such deal was a $200 million contract within the FBI's Criminal Justice Information Services Division. Headquarters for this division was in Clarksburg, West Virginia, a four-hour drive west of D.C., and after a lengthy proposal process, we were one of three bidders to come and negotiate terms in person. The contract incumbent was Lockheed Martin, and we were all psyched to outdo one of the government's top contractors.

I was part of a four-person negotiation team that included Ben and Bill Ballhaus, our company's new CEO. We all arrived the night before, gathered for breakfast, and then made our way to the FBI offices, where we were ushered into a large conference room for our meeting.

There were about twenty FBI officials and lawyers waiting for us, and as we entered I could hear the *oohs* and *aahs*. Everyone's eyes were on Romeo. Before we took our seats, I leaned over and whispered to Ben, "This is going to go well."

It got better. I was seated next to an FBI employee who was eager to tell me he was a Lions Club member, the main sponsor of Leader Dogs for the Blind in Michigan. I was proud to tell him how Romeo was my third guide dog from Leader Dogs, and how the Knoxville Chapter of the Lions Club had changed my life with its generous support.

Our bid was a good one, and we deserved to win, but I told Ben afterward that Romeo deserved a commission check. In a room full of hard-nosed negotiators, Romeo had clearly softened up the room and made them that much more receptive to what we had to say. Anytime you can take a guide dog into a contract negotiation, it instantly injects humanity into the room. No longer were we businessmen and -women or attorneys; we were all just people. Take advantage of everything you have at your disposal. Sometimes your disadvantages can be an advantage.

I loved the work I was doing. Each new bid was like a fresh puzzle to solve, a Rubik's Cube with a unique code to crack. Optimizing the numbers was only the start. They had to be presented in a way that made sense to the team, so we could persuade the operating units that some sacrifice in one area of the bid would pay off in another area.

It didn't take long for people at SRA to recognize that we were winning about 40 percent of the deals I worked on, while the win rate for most other bids was around 20 percent. Ben was appreciative, and I was rewarded with promotions and raises. He sang my praises to others, both inside the company and outside, as I soon learned.

One day, I received an inquiry from a job recruiter with Hewlett Packard. I'd gotten nibbles from headhunters like this all the time, and I usually ignored them. But this approach was different. I was put in contact with two people at Hewlett Packard who wanted me for the job. They had known Ben for years and had worked with him at one point. They'd seen him recently, and evidently Ben had been talking me up. Now they wanted to poach me from Ben, their friend. They pursued me as I'd never been pursued before.

They asked me what I was making at SRA. A bit of advice: never tell anyone the answer to that question. Don't even lie about it. Because when someone wants to make you a job offer, it doesn't matter what your current employer is paying you. All that matters is the number that would make it worth your while to switch. How much money would you need to leave a job where you're already happy?

I knew that my compensation at SRA was within the normal range of what the market was paying for jobs like mine at the time, but it wasn't at the top of that range. So, I made a pretty bold request. I gave HP a number 40 percent higher than my salary at SRA. I told them, "This is what I think is good market compensation for what I do."

A few days later, without even conducting an in-person interview with me, HP sent me a formal job offer by email. My proposed starting salary was the exact number I'd given them.

Whoa! Now what?

SIX

DARE
TO BE GREAT

"OH MY GOSH, ARE YOU GOING TO TAKE IT?"

Evie was excited about the HP offer.

"I think so," I told her over the phone. "I can't believe they met my number."

"They want you so badly, and SRA can't match that salary, can they?"

"We're about to find out."

It was a little shocking to get such a generous job offer so suddenly out of the blue.

Now came the tough part. I needed to have a difficult conversation with Ben—my boss, my mentor, my role model. I felt uneasy about going in and telling him in not so many words that he'd lose me if he couldn't meet the HP offer. And he was sure to feel betrayed once I'd told him that his former colleagues—people he'd considered friends—had tried to poach me after he'd talked me up to them. This was going to be a difficult conversation, and I wanted to make sure I handled it appropriately.

I also knew that no one ever recognizes your true worth until you recognize it yourself. I'd created a lot of upside for SRA in my years there, and not just because I was good at my job. I helped stand up SRA's entire new practice area in managed services, involving myself in details such as staffing and writing service-level agreements. I wasn't afraid of going outside the pricing swim lane. I involved myself in every facet of our products to ensure that we always put our best foot forward in everything we did.

For example, we were participants in a big contract with multiple partners, where SRA held most of the work. The customer wanted all the partners to lower their prices as a condition of renewing the contract. But I'd researched this area thoroughly, and I knew that it wasn't a very competitive situation. There was no one else out there positioned to do the work as well as us, in the way we were doing it. So, we really had no motivation to reduce our prices.

Unfortunately, the line-of-business executive in charge of the contract saw it differently. He wasn't comfortable with telling the customer, "No." What if the customer got mad and dropped us? The executive said he'd rather we lower our prices than take that risk.

It took some time and effort, but I ended up changing his mind. I showed him the numbers and explained how we held a strong advantage that would be unwise to give up.

Most of our partners on this deal complied with the customer's wishes and dropped their prices, but we kept ours flat. It ended up being the right decision. We won the contract and maintained the pricing that left us with a solid rate of return.

That was just one instance where I was willing to get outside of my comfort zone. As a deal strategist, I nearly always had to apply downward pressure on pricing to keep us competitive, so that was my natural instinct. This time, however, the data and analysis pointed in a different direction. I probably earned the company fifty times my annual salary on that one case alone. There was no question in my mind that SRA would be smart to pay the price of keeping me around.

My overriding concern was that I didn't want to offend Ben. He'd stuck his neck out for me many times in the past, especially when I started at SRA, and I didn't want to appear ungrateful. I'd do anything for Ben, and I'd proven it many times. The truth was that I didn't want to leave SRA. I was constantly being groomed by all the whip-smart people around me. Ben and I had a great, easy, back-and-forth rapport. Like all good bosses, he knew what he didn't know.

When I'd go in and ask Ben for advice on a job I was working on, he knew not to tell me what to do. Our deals were all so complex that there were few right or wrong answers to dispense. Ben would never say to me, "Here's what you should do." He would say, "Tell me about what's going on." Then he'd ask me, "Have you thought about this?" "Have you thought about that?" "What do you think?"

With his questions, he'd guide me down a path of inquiry where he might express his opinions and preferences, but at the end of the day, he let me stand on my own, because he trusted my judgment and my understanding of the job.

On a Wednesday morning, I called Ben and said: "Hey, I was approached by HP. They're really excited, and here's what they've offered me." I forwarded Ben the offer. He recognized the names on the offer and grumbled about his friends' involvement. He felt disrespected by them.

I said to Ben: "You know, there's a big opportunity here for me to make more money, but I don't want to leave. I love working for you, I love working for SRA. But that's a big difference in compensation, and I have a responsibility to my family, too. What can we do to bridge the gap? Should I give up on the idea of staying here? I definitely want to stay, but this is a lot of money for me to pass up."

Ben had questions. Had I told HP my current salary at SRA? I assured him that I hadn't, and Ben appreciated that. Then he asked me how HP had arrived at the number they'd offered me.

I told him the number was mine. "That's what I believe I'm worth," I said.

Ben took a long pause. No company hands out 40 percent raises every day, so I knew I was putting him on the spot.

"I'll go talk to Pat," he finally said. Pat was his boss. Ben thanked me for coming to him, and we hung up.

I waited two days to learn my fate. Ben called to tell me the verdict. "Here's what we're able to do," he said. "This is not a 'raise.' It's a 'market correction.'" He emailed me SRA's counteroffer. My new salary at SRA *exceeded* the HP offer by more than 5 percent.

"We want to keep you," Ben said. "We love having you on the team. We're giving you more than what they've offered you to show you how much we appreciate what you've done for the company. We want you to keep doing it for a long time."

I accepted Ben's counter on the spot and thanked him. My next call was to Evie right away to give her the good news. Our vacation plans would get an upgrade.

The lesson I learned from this experience is that you can't go wrong by daring to be great. Give it your all, help your employer succeed, help your partners succeed, and there's no limit to your success. You will be recognized and rewarded, even if the route to recognition is not always direct.

This little episode also reinforced my belief that you can ask for anything and everything you want, as long as you use the right words, delivered with the right tone. My approach to Ben was one of humble, nonconfrontational curiosity and in the spirit of our years-long partnership: Can you help me solve this problem? I used a little humor, too, to relieve some of the natural tension surrounding the situation. The last thing I wanted to do was sound like I was making a demand for something I felt entitled to. That would have made it very easy for Pat, Ben's boss, to say: "Who does Chad think he is? Go tell him to take a hike."

The stark truth is that we rarely get what we want in life if we don't ask for it. That's why it makes so much sense to be very careful about how you ask. You always want to ask in a way that makes the other person *want* to do it for you.

DEEP INTO STRATEGY

It's a thrill to be paid what you're worth, but it's also a little nerve-wracking. I'd just gotten a raise of almost 50 percent. How do I go in tomorrow and be 50 percent more valuable? I didn't think about it in quite such stark terms, but I did feel like I was on the spot. Now that they'd given me this huge raise, I'd better be ready to show that I'm worth it.

That didn't prove to be a problem. For a gigantic, complex $5 billion contract with the Army, I ended up carrying a workload that I found out was shared by four different people at other companies. I devised the financial strategy, set the price target, and then coordinated the work with more than fifty different subcontractors or "teaming partners." There were four separate proposals within the bid, covering a total of five hundred labor categories—all requiring sealed packages that we had to collect and deliver to the government. To hit one of our deadlines, I worked twenty-three hours straight.

Our whole team was stretched to its limit and beyond by that bid. After we won it, the infamous Army contract became the only deal we commemorated annually with a walk down memory lane with photos and stories. Trauma has a way of pulling teams closer together. The Army deal hardened us all for the next deal, and the one after that.

Political changes had caused sudden drops in the appropriations from Congress around 2012, and the pressure to compete grew more intense. In one big successor contract bid, the government agency gave us a huge spreadsheet made up of thousands of input cells, with many of them read-only protected. It was impossible to perform a quick copy and paste across the entire range. The tedious process of copying and pasting our proposed formulas into each one of the unprotected cells would have taken days to do manually, and we didn't have that much time—especially when the customer would give us an updated pricing template to use every so often.

In this case, my blindness created a real advantage. I have a very deep knowledge of what are called back-end Excel libraries, which allow you to automate all kinds of Excel processes. Think of these libraries as the instruction manual that is used when Excel needs to talk to another program, like Microsoft Word or Microsoft PowerPoint during a copy-and-paste operation. Most Excel users have little or no knowledge of the libraries, but for me, they are a daily lifeline. So, I wrote up a simple automation tool that would go in and automatically update the government-provided spreadsheet with our bid proposal data. It took about two minutes to complete a job that would have otherwise taken days. I don't know if any of our competitors knew how to do that, but I do know that we won the contract.

As my coding skills continued to sharpen, my off-hours consulting work really hit its stride—the result of a series of unexpected circumstances.

Back in 2007, a former colleague from Accenture had contacted me out of the blue because one of his client companies had a blind employee who needed to work with software for customer relationship management, or CRM. The CRM software was called Siebel, made by the software giant Oracle, and no one knew how to make it work with JAWS. I didn't know anything about Siebel, but I promised I'd take a look.

It was a complete mystery why the two applications couldn't talk to each other, which left me to poke around at Siebel's program architecture, searching for various back doors and workarounds. It took months of trial and error to find a solution, which required me to write tens of thousands of lines of computer code. It was interesting work that paid well, and I didn't think much of it after that.

Then, a few years later, a brief exchange on an accessibility bulletin board led me to a conference call with an Oracle technical account manager and a software specialist from Freedom

Scientific, the maker of JAWS. For years, they'd assumed there was no way to make Siebel a JAWS-compatible application. Both were astonished I'd found a fix and wanted to learn more about what I'd done. After that call, Oracle began sending prospective clients my way.

Among those clients was a large Canadian bank that ended up hiring me for several different big projects spanning about four years. Under Canada's strict disabled-access laws, the bank was eager to make all its Siebel-related applications completely accessible to the blind, in both English and French. They wanted all the work fully documented and they wanted to own the source code.

I knew from my experience that if you're the only person on the planet with a solution someone needs, you can write your own ticket when it comes to pricing. The fees I earned from the bank for the hundreds of hours I put in over those four years helped finance our college savings accounts, our retirement accounts, and a second, much nicer investment property in Florida.

If there's a moral to the story, I'd say that whatever you're good at, try to take it to its limits and then beyond. Try doing the thing in your field that everyone says can't be done. Maybe you succeed, maybe you fail, but you'll grow like crazy, and there's no telling what might come of it. Dare to be great. If you have faith in your gifts, the rewards are out there. Choose a daunting challenge for yourself, and then, to quote my high school wrestling coach once again, just grab it and growl.

One day, I learned I'd be getting a new boss because Ben had earned a promotion and was now the vice president in charge of the business line covering civilian agencies. I was happy for Ben, and although I'd be losing him as a boss, I was gaining a great new partner. Convincing the executives in charge of lines of business to go along with the pricing recommendations is very often the most difficult chore in the process. Not with Ben. He not only knew how the game was played, he'd even written some of its informal rules. He'd schooled me in my strategic approaches, which is partly why he had confidence in my recommendations. On bids with the Securities and Exchange Commission and with the Federal Energy Regulatory Commission, we worked side by side, driving our strategy through the entire proposal life cycle, and won both bids, which created hundreds of millions of dollars in revenue for SRA.

It wasn't long after I got my big salary increase that we realized the best use of my time was not working on one deal at a time. Only about 15 percent of my time was spent on the strategic research and positioning that I was really good at. The other 85 percent was spent on tasks that most other talented professionals at SRA could do just as well as me, if not better.

So I shifted into spending almost all my time and attention on research and strategy for numerous deals, all at once, while two or three other pricing analysts took up a lot of the work I'd done myself until recently. And I found that the deeper I could go into strategy, the more I could achieve new insights, create new financial and staffing models and tools, while attaining new levels of mastery in the domain. My technical understanding of our work was so refined that I'd estimate how many people we needed to do the work by task area. Even for deals not assigned to me, engineering teams would seek out my help on projects to estimate and justify projected staffing levels.

I would also research bid documents all the way through to the future congressional appropriations affecting the bid, all to predict where work would be several years down the road. I'd be looking at a $400 million bid that would frighten the business line manager because our first-year margins were razor thin. And I could say: "Here's my plan. Considering I know where we play well in this space, the statement of work they've asked for is ugly on paper, but I've got a plan for us over the next five years to follow the appropriations dollars and improve this profit-and-loss statement to something that's much more palatable."

My big raise offers a second lesson here. Once I was making more money, the higher-ups at SRA more readily recognized that I was being sorely underutilized. So, the market correction in my salary, prompted by the HP offer, wasn't just a big personal financial boost. It raised my profile at the corporate level. It set the stage for the next phase of my career.

THE BIG ASK

In September 2014, the University of Tennessee at Knoxville honored me with their Accomplished Alumni Award. But more important, during the ceremony, they also awarded my mother an honorary Accomplished Alumni Award, even though she'd never gone to UT, or any college for that matter.

Three years earlier, Evie had suggested that I reach out to UT's alumni office on behalf of my mother. We wanted her to get the recognition she deserved for her hard work and dedication in helping me get my degree. In my first call to the office, I'd asked about how we could grant her an honorary bachelor's degree from UT, and that's still the goal. After all, she'd read every word of every chapter that was assigned to me while in business school. How many of my fellow graduates could make the same claim? I imagined precious few had read every word of text as carefully as my mother had.

The more I told the alumni office about my work and my job history, the people there became convinced that Mom and I should both be honored at a special luncheon ceremony on campus. The ceremony would be announced to honor me only. We agreed not to tell my mother in advance that she was also going to get an award.

I gave a brief speech at the luncheon where I credited my mother for "being my eyes" during my time at UT. I told the crowd: "She painstakingly read aloud and recorded every page of telephone-book-sized business management books for me, sometimes until laryngitis forced her to stop. Often not having slept a wink, she would be at her bookkeeping job by 8:00 a.m. My mother's sacrifice was a big part of my success." And then I surprised her with her honorary alumni award. I wish I could have seen the look on her face, but I could hear how surprised and happy she was.

I had told Joe, my new boss, about the award because I needed a few days off to fly to Knoxville for the event. Later, I forwarded Joe a link to a write-up on the alumni association website: "Business Alum Overcomes Blinding Disease, Receives Accomplished Alumni Award."

A few days later, Joe called to let me know he'd read the alumni article. "Chad, you've got a big story here," he said. "What can we do at this company to help you take the next step?"

What an interesting idea.

I'll never turn down an invitation for self-development. I'm always looking for an edge, for a way that I can raise my game. I always want to be just a little bit better version of myself. So, if I can learn something new or join a new network that can help me maximize my potential, yes, sign me up.

But I'd already done any number of run-of-the-mill leadership training courses, and now I was eager to challenge myself with something much more meaningful. Joe and I both shared a common belief that when you get a chance to step to the plate, always swing for the fences. That's how we'd taken SRA through a tremendous phase of growth in a shrinking marketplace. If you dare to be great and swing for the fences, you won't connect every single time, but when you do, you'll hit some home runs. I figured Joe shouldn't be surprised if I come back to him with a Big Ask.

I'd long put off getting an MBA, and I toyed with the idea of asking Joe for tuition support toward that. When we lived in the D.C. area, I briefly considered applying for the Wharton MBA for Executives program at the University of Pennsylvania, but the time demands would have been impossible. The program would have required me to spend every other weekend for two full years in Philadelphia, a three-hour train ride from home. With a small child in the house, it seemed ridiculous to be away from home that often. You can never get back those early years with your children, and my job at SRA was already very time consuming. Working on deals is an around-the-clock affair. The customer deadline drives the project work schedules, with no regard for the hours in the days, nights, weekends, and holidays.

Now I turned my attention to Harvard. I'd always wanted to go to Harvard, but the Harvard MBA program requires you to take off from work and live on campus for two years. That wasn't an option for us either. However, my research turned up an alternative, the Harvard Program for Leadership Development, promoted as a "unique learning experience specifically designed to fast-track your career." The program involved about a year's worth of remote-learning coursework and three immersive modules held on-site at the Boston campus, two weeks at a time.

It seemed pretty doable. The faculty, administration, and coursework are world-class, and so are the networking opportunities. I could swing six weeks away from home spread out over more than a year. And the target market sounded exactly like me: "specialists and star contributors with at least 10 to 15 years of work experience who have been identified as outstanding prospects for increased leadership responsibilities."

So I came back to Joe with my Big Ask: send me to Harvard. The price tag was pretty high—roughly the cost of a year of Harvard tuition, room, and board.

What you need to know about Joe is that he is one of the smartest people I've met. And he's also an aggressive dude and a fierce competitor. But the cost of sending me to Harvard left him just a little bit uncertain. The only time he could ever recall the company doing something similar was for an executive being groomed to become chief financial officer.

"Chad, I love the way you think, but I don't know if we can do this," he cautioned me. "I can't promise you anything other than this. I'll sell it. I'll sell my ass off for you."

I eagerly waited to hear the news for the next several weeks. It felt both thrilling and nerve-racking at the same time. I was initially fearful because what would happen if the company didn't approve it, and I went from being this A-player to getting denied? No doubt this would damage my ego, and it would get around that I'd asked for something outrageous and been denied.

Then I got the good word from Joe: Bill, the CEO of SRA, had personally approved the funds. I was ecstatic. But I couldn't tell anyone. Because now I had something else to worry about.

What if I applied to Harvard's program and I didn't get in? I'd look like an idiot, someone who asked the company to pay for something extravagant that I wasn't even qualified for.

And that's often the challenge when you dare to be great. What you're really doing is putting yourself out there and risking failure, possibly looking like an idiot for trying. Vulnerability was never one of my towering strengths. And remember, I care how I look. But I care more about reaching my full potential. Harvard had been a longtime dream of mine since childhood, so I was prepared to deal with the damage to my ego if it didn't work out.

Luckily, the cards fell favorably for me. In April, I was approved for the Fall 2015 class of the Harvard Business School Program for Leadership Development.

HELLO HARVARD

My coursework at Harvard began with remote-learning coursework on accounting and analytical skills all through the month of September. Then, in October, the entire class of 170 business executives converged for two solid weeks of sessions at the business school campus in Boston, just across the river from Harvard's main campus in Cambridge.

The Harvard name had such cachet for me that I had very lofty expectations when I first entered the program. As high as I'd set the bar, I wasn't disappointed in the least. Never before had I been in a place that had put so much advance thought and consideration into what I would need.

I'd arrived a day early to acclimate myself and Romeo to our surroundings. I was introduced to a senior facility manager, who showed me around campus and led me to where I'd be staying for the two-week session. She gave me her cell number and the number of the concierge at the dining hall, which was really more like a fine restaurant. She told me I should call either number if I ever needed anything, any time of day or night.

I was the very first blind student this particular Harvard program had ever admitted, so they brought in a disability consultant to ensure that all the coursework had been made accessible for me. Much of our class time would be spent discussing case studies that we'd read up on in advance. The materials in the case studies sometimes contained images of graphs and tables that had all been transcribed into text for me. All the financial spreadsheets related to the case studies were presented in Excel, the format I'd told them I preferred. I had everything else I needed in electronic form, and all my software accessibility tools worked flawlessly with Harvard's systems.

It was ridiculous. That was the only word for it. Never in my life had I ever felt so cared for outside my own home. One morning, I was at the gym and discovered I was running late. I still had to shower and get dressed for class, and there was no way I'd also have time to make it to the dining hall for breakfast. So, I called the concierge number. "Look, I'm running late for class," I said politely. "Could you possibly send some breakfast to my room?" By the time I made it back from the gym, there was a full spread of hot breakfast waiting for me in my room.

It was next-level service, in every respect. In the dining halls, a server was dispatched to take my order and then prepare a plate for me from the buffet.

It was five-star service at a five-star school.

I found the classwork was so much more interesting and engaging than when I'd been in school fifteen years earlier. Not only was I more mature now, with stronger business and computer skills than when I was in college, but the assistive technology had also vastly improved from the relatively clumsy software I once knew. For the first time in a classroom, I felt that the tools I'd been given weren't getting in the way of learning the material.

But the biggest difference was my fellow students. When you're in a class discussion with that many smart, accomplished people, everything that everybody has to say is interesting. Even when I didn't agree with what I was hearing, I was thinking about how I might be wrong. My classmates were all a very special kind of smart, too. All of them were younger executives who had enough humility to know they had more to learn by coming back to school. There wasn't one show-off in the group. It was as though everyone had checked their egos at the door. They were humble and confident enough to know that you learn more listening than talking, and that everyone offers something you can learn from.

One of the most inspiring professors I had was Bill George, the former CEO of Medtronic and now currently a full-time professor senior fellow at Harvard Business School. As the best-selling author of *Find Your True North*, his message about the importance of finding one's purpose and passion in life resonated deeply with me. He cites the phenomenon of PTG—post-traumatic growth—in which terrible events can be used as crucibles to "transform your wound into a pearl."

In *Find Your True North*, Bill pointed out how Oprah Winfrey had turned the torments of her youth into a source of personal strength. "Given the abuse and poverty she experienced earlier in her life, it would have been easy for Winfrey to feel like a victim," Bill wrote. "Yet she rose above her difficulties by reframing her story in positive terms: first by taking responsibility for her life, then in recognizing her mission to empower others to take responsibility for theirs."

That notion of using negative events to create positive results rang especially true for me. I talked it over at length with one of my classmates, Greg Jarząbek, whose experience of losing his mother to cancer motivated him to found Trustedoctor, an online startup that coordinates patient care with top doctors and hospitals.

Our weeks together felt like a learning laboratory. It was an adventurous place where we could explore ideas together without anyone worrying about being judged. Many of the case studies didn't offer any right or wrong answers. They'd been written up to offer us all kinds of nuanced shades of gray. Business leadership is all about making decisions without enough information to know which course of action is objectively better than the other. You face hard choices, without a certain path to getting to your desired result. The discussion was all about how to assess the available information in ways that are most helpful to arriving at the best decision possible.

True to my nature, I wouldn't make a comment unless I really felt it would drive the dialogue forward in a meaningful way. I always sat in the back of the class with Romeo, so on the occasions when I did speak up, every head would have to turn. And then what I had to say often shifted the class conversation in some unexpected ways. A bunch of people told me they were just blown away by my contributions. I think a lot of the international students were especially impressed because they'd never met anyone like me. Outside of the United States and Europe, blind people are rarely in the workforce at all, much less at the executive level.

For each graduating class in this program, the students vote on one person as the speaker for the graduation dinner. As we neared the end of Module Two, I got the distinct sense that my classmates were going to vote for me as the speaker—the only person in class incapable of reading a speech from prepared notes! I couldn't be certain, but I just had a feeling it was meant to be.

When I returned home to Atlanta after the Module Two class-work had concluded, I started wrestling with the idea of being chosen as the speaker. I decided I needed to prepare for that possibility, rather than risk being unprepared. I thought of Peyton Manning, my fellow UT alum, and Super Bowl MVP and champion, who famously said, "I've never left the field saying, 'I could have done more to get ready.' And that gives me peace of mind." I wanted that peace of mind. And although we were learning about public speaking at Harvard, I was sure I needed one-on-one help with writing the speech itself. I was taking a shower one Sunday morning while listening to the popular *Smart People Podcast*. The guest that week was a public-speaking coach and author named Garrison Wynn. He was so smart and funny (he'd started out as a stand-up comedian) that I decided I needed his help to develop my speech. It felt serendipitous to hear him expand on the keys to great storytelling that morning.

I hopped out of the shower and raced to my computer before I had a chance to forget his name. I sent Garrison an invite to connect on LinkedIn, as cold as an ice cube, with a note saying I had a unique story and needed some help telling it. I included a link to the UT article about my alumni award.

To my amazement, Garrison responded the next day, and we arranged for a phone call later that week. He explained to me that he gets frequent requests for personal coaching, and he rarely has any time to spare because he gives more than a hundred talks per year, and most people who inquire are either lacking in talent or don't have a message that stands out.

"But you," he said. "You've got a real story." He was willing to make an exception in my case. We made arrangements to spend some time on the phone and then get together in Houston, where Garrison's office is located and where my cousins live.

I flew into Houston on the Friday before my fortieth birthday. My cousin Jeremy met me and Romeo at the airport, and we spent that Friday and Saturday catching up with the rest of the family. That included Jeremy's older brother, Mark, who had visited me in my South Knoxville house back when I was twenty-one and had cautioned me about the company I was keeping. Although my birthday wasn't until Monday, they threw a party for me on Saturday night.

Jeremy and I had been like brothers growing up, and he was a great guide and companion the whole weekend. Early Sunday morning, Jeremy and I showed up for my appointment at Garrison's place, where the three of us spent hours grinding out what was only a twelve- to fifteen-minute speech. We took every word, every sentence, very seriously—carefully scrutinizing the meaning of each word while considering the alternatives. It was crucial that the words connect with everyone, Garrison explained, because if an audience member has to stop and think about what I've just said, then that person is lost for the remainder of the talk. Outstanding speakers, he said, are able to tell stories that simplify ideas, while sowing seeds of emotional surprise at every turn—leaving the audience engaged and curious about what's coming next.

Garrison also cautioned that many speakers who discuss life-changing traumas like mine risk leaving the audience feeling sorry for them, which was the last thing I'd want to do. It is also imperative that the talk use lots of humor. My story is a heavy dose of reality, and for people to hear me in the way I intend, I have to use humor to lighten the load. That way, the audience can take in my message and walk away from my talk feeling better than before.

As the day dragged on, it didn't seem like we'd gotten very far. There was so much more to this than I had imagined, and I grew concerned we wouldn't be able to finish in time.

"It doesn't look like we'll have enough time," I finally said to Garrison at one point.

"Don't worry, we'll get it done. We'll keep going until it's done," he said. "I give you my word."

We spent fifteen hours together that day and didn't finish until after 11:00 p.m. It took us more than an hour to get back to Jeremy's house, and we slept barely three hours before it was time to get back in the car to make my 6:00 a.m. flight. I spent the morning of my fortieth birthday waiting for a flight home to Atlanta at the Houston airport.

Later that week, our class convened again in Boston for our final two-week session. As I'd anticipated, I was nominated and elected to give the graduation speech at the end of the session. I was prepared to rise to the occasion.

"YOU REALLY GAVE ME A GIFT"

Everyone was relaxed and having a good time at the Harvard Club except me.

About 250 people had gathered for the graduation banquet, and all I could do was keep playing back my speech in my earpiece, practicing it silently while everyone around me was chatting.

Although we'd written the talk about two weeks earlier, I hadn't had much time to commit it to memory, and reading from notes was naturally not an option. I'd clocked the speech at just over twelve minutes, and I didn't want to rush through it. I was feeling nervous, although I might have felt even more nervous if I could see the crowd.

Mike, the professor who led our class, asked me if I was ready. It was time. He took to the podium and introduced me, which drew an instant roar of cheers and applause.

I began my talk by expressing my gratitude. I thanked the faculty and program staff, asking everyone to give them a round of applause. I thanked my classmates for their friendship and their honesty through some very open and sincere classroom discussions. And then I expressed my personal gratitude for living in a time period that offers me so many tools for learning and navigating through the day.

"Like I've always said, now's a great time to go blind," I told them. "Not that I'm recruiting, but just twenty or thirty years ago was not a good time. Thirty years ago was a terrible time. Trust me, if you're looking into going blind, now's the time."

Big laughs and applause. Garrison had coached me on that line, and it killed.

I went on to tell the stories of my misadventures when my eyesight was poor: Falling through the missing stairs, tumbling into the roadside ditch, and how my blindness forced me to flush years of college down the drain. Then I told of how my belief that "excuses are for losers" rescued me from those dark days and set me on a more productive path.

I said that when I got to Leader Dogs for the Blind, the courage of my fellow classmates "taught me life's most valuable lesson.

Happiness is not a feeling. Happiness is not an emotion. Happiness is a decision that each of us makes every day when we wake up. We either choose to deliberately frame our perception, or we allow the circumstances of life to determine our happiness. Because this, ladies and gentlemen, this is a key ingredient for improving your happiness and your vision. . . . I have found from my own personal journey that when you choose happiness, success ultimately follows."

I mentioned how studies have proven that our levels of both happiness and suffering depend on the stories we choose to believe about our experiences. And then I spoke about my own experiences, taking physical risks, like getting thrown across a bar floor by a mechanical bull. That got more laughs.

The room grew hushed when I turned toward the conclusion. "Ambition is important. You have to weigh the risks and rewards. But I refuse to live my life in fear. And I refuse to live my life according to other people's expectations of what they think I can do. And I refuse to let my lack of eyesight limit my vision. You may call it blind ambition, but I call it life. We all have obstacles. What obstacles are getting in your way that you need to overcome to improve your vision and make a difference in the world we live in? Thank you."

I wasn't ready for the explosive applause and instant standing ovation that followed, accompanied by plenty of hooting and hollering. Mike, the professor who'd introduced me, shouted through the racket, "Oh, my God, that was incredible!"

My only fellow classmate from Atlanta, Tomer Zvulun, the general and artistic director of the Atlanta Opera, told me he could "hear music" while I was speaking.

A small crowd formed around me afterward, and I heard members of my living group saying, "Chad! You killed it!" and

"Way to go!" More than a few told me I should be a professional speaker.

Then one classmate pulled me close and mentioned his recent divorce. "I'm going through a tough time now," he said. "And, boy, you really gave me a gift. Thank you."

Another classmate, whom I think of often to this day, came up to tell me about his young daughter whom he'd lost to cancer. Something that I'd said had given him hope for seeing her loss in a different light. At the mention of his daughter's name, he broke down crying in my arms.

In that moment, I felt overwhelmed that my simple words could provide comfort to a man whose loss was so much greater than my own. A feeling instantly electrified my body. I could feel it coursing through my veins. I'd never experienced anything like this before. At that moment, I knew I was doing what I was meant to do. What I was uniquely qualified to do. To this day, I tear up at the memory and the sense of responsibility it aroused in me, that it was important for me to share my story.

I was never a warm and fuzzy person to begin with, and I'd been hardened with a tough-love upbringing and all the challenges presented by my condition. And although I don't consider myself a very emotional guy, the entire experience of that night left me feeling both overwhelmed and fulfilled in a way that's hard to put into words.

From time to time, people had told me that I inspired them. I thought that was nice of them to say, but I never really took them seriously. The Harvard talk showed me how powerfully people could be moved by my story if I told it with humor and addressed our common human condition. There's something deeply satisfying about using your biggest obstacle as an opportunity to inspire others to higher dreams and aspirations. Helping people recognize their own capacity to grow actually makes going blind worth it. And that was just my first attempt!

I thought of my professor, Bill George, and what he'd said about purpose and passion. What if I took this responsibility seriously? How good could I be? How many people could I reach? How much of a difference could I make? Could this be the crucible that transforms my wound into a pearl? My biggest burden—my blindness—could become my biggest gift to others.

I felt excited and terrified at the same time. Because it meant that I'd need to go through the entire nail-biting ordeal of public speaking again and again.

SEVEN

OBSTACLES
EQUAL GROWTH

"STEP LEFT. LEFT AGAIN. NOW STRAIGHTEN OUT."

It's the voice of Paul, my friend and ski guide for the day, getting me oriented for our run together down Cirque Headwall in Aspen, Colorado.

"Okay," says Paul. "Ready when you are." We push off, with Paul behind me, connected via a wireless headset.

A lot of skiers get scared at the top of Cirque, which is among the steepest expert-grade slopes at Snowmass ski resort, rated "Double Black Diamond." I don't get scared on the expert slopes, because I can't see what the big deal is all about.

"Right turn," Paul says in a calm, steady tone. "Steady. Hard left. Hard right."

It's a thrill to move this fast through the cold, thin mountain air. The wind is whipping into my face, and the snow is crunching beneath my skis. There is also something about the stomach-churning sensation of making a turn on the ski edge, and sharply carving across the slope that feels like I'm being shot out of a canon.

I started skiing in 2014 at the relatively advanced age of thirty-eight. I wish I'd started years earlier because there aren't many things I enjoy more. There's the feeling of growth, of overcoming one obstacle after another. And there's the teamwork. Downhill skiing without eyesight depends upon constant communication between my ski guide and me through our matching Bluetooth headsets.

Sonia was my first ski guide, a former professional skier who has been specially trained to teach skiing to the blind. On every run when I started, my high school friend Paul skied behind Sonia and me, with a video recorder on his helmet, and also connected via Bluetooth. By studying Sonia's commands on those early runs, Paul learned to be my ski guide, so now he and I often go out together, just the two of us.

I trust Paul and Sonia like few other people in my life. If you've ever done a trust fall exercise—closing your eyes and falling backward into someone's arms—every ski run for me is like a twenty-minute, forty-mile-per-hour trust fall. Sonia has told me that my blindness gives me a big advantage in developing my skiing skills, because I never get distracted or frightened by the terrain around us. I stay in the moment, turn by turn, because I have no choice. That gives me a huge advantage over sighted skiers, who usually require many more years of experience than I have before they dare take on the steepest slopes.

I always try to make the most of my limited time on the slopes, always pushing harder and harder. I feel like, if I'm not falling down, I'm not pushing hard enough. I want to stay at the very edge of my abilities at all times, so that I know I'm stretching and growing.

Skier slang for a major wreck is a "yard sale" because skis, gloves, and poles wind up strewn across the snow when you go down hard. I expect to wreck, but I've only rarely been hurt. On the last day of my first ski trip, I injured my shoulder pretty badly, but didn't find out it had been dislocated until after I'd skied two more runs. Another time, on a very steep expert-level slope, I fell fifty feet down into a "tree well"—the open area at the base of a big evergreen, sheltered by tree branches, where not as much snow accumulates.

On the way down into the tree well, I just kept hitting one branch after another and had no idea what happened until Sonia arrived to guide me out. People have died in tree wells because the snow there is such fine powder that you can sink into it like quicksand while trying to climb out. When Sonia called the ski patrol for help, I was hoping I'd get pulled out by a snowmobile with a winch. No such luck. Instead, a kid from the ski patrol gave me one of his poles to help me climb out, pulling myself upward six inches at a time. It took me an hour, and by then I was so exhausted that I had to take a rest break, which I never do during a day of skiing.

Lots of blind people ski, and visually impaired skiing is a Paralympic event, but when I tell people that we go skiing in Aspen, they sometimes stare as though I were a monkey at a magic show. We can never let other people's opinions define what we can or cannot do, but the truth is that sometimes other people's ignorance can make it very difficult.

Evie and I were visiting the jungle in Belize with my cousin Jeremy and his wife, Amber, when we decided to take a two-hour drive to a commercial zipline that had been strung some two hundred feet above the jungle floor. As we paid for our tickets, I told the zipline operators that I might need some assistance getting on and off the lines between each platform because I'm blind. They suddenly got nervous and refused to let me go. For almost twenty minutes, we couldn't persuade them that there would be no problem in letting me ride the zipline.

Finally, I just turned my back and started walking toward the zipline platform. Evie, Jeremy, and Amber took my cue and followed. That was all it took for the zipline operators to lose their resolve and let me go.

Here's the funny thing. Some of the people ahead of us in line became utterly terrified when it was time for them to step up to the platform. They freaked out at how high up they were. And guess what the zipline operators told them to do?

"Just close your eyes!"

I thought that was pretty ironic. When it was my turn to hop on the launchpad, I turned to the operator and asked him, "Should I close my eyes?"

All my life I've been a smart aleck, and since I've been blind, I do miss seeing the reactions on people's faces.

Skiing is all about managing risk, just like life, so it tunes your mind to be aware of unnecessary risks. One day, we were racing down the mountainside in whiteout conditions when our speed and the high winds caused a crackling in my earpiece. I couldn't be certain of the next direction being called, so I executed an immediate hockey stop. A tingling in my spine had told me to stop before I even had time to assess the danger. We regrouped and waited a few minutes for the wind to calm down. That kind of fear, the fear of unacceptable risk, is healthy fear. It keeps us alive long enough to manage all the acceptable risks that enrich our lives.

Often when we're afraid to do something new, we're really avoiding an acceptable risk that will help us feel alive and grow. No one wants to get hurt, but avoiding what you fear won't protect you. You can get hurt doing just about anything. Once when I was visiting home in Knoxville, my father drove me to his local gym and convinced me to leave Romeo back at their house. Then, while leading me by the arm, he absentmindedly walked me right into a wall. The collision sounded like a watermelon hitting the ground, and he had to take me to the hospital for stitches. It was a worse injury than any I'd suffered on the ski slopes.

One of the many lessons I've learned from downhill skiing is that surface-level fears—such as extreme elevation or steep terrain—are just that, surface level. We're often better off when we are not encumbered by fears and simply face life's steep terrain at full speed, always outside our comfort zones. I can live with the occasional crash at the edge of my abilities, but I cannot live with playing small and safe. I'd never claim that everyone needs to go downhill skiing or ziplining. But I do believe that everyone should spend more time moving past the edges of their comfort zones and getting comfortable with the discomfort this entails.

The most significant factor affecting performance of any kind is the positive belief that when you stretch your abilities, you will prove yourself far more capable than you had thought possible. I believe this, because my entire life is an ongoing experiment in what you can achieve if you just believe you can.

If you set your goals far higher than what you think is attainable, I guarantee you'll be surprised by how far you'll get. The discipline is not that different from what it takes to ski blind. Get a coach. Work your skills. Find your edge.

HOMAGE TO ROMEO

My Harvard speech marked the start of six months of transitions for our family: a new job for me, a new baby, a new guide dog, and the passing of our old and trusted friend Romeo.

I'll start by paying homage to Romeo, who died in March 2016. Romeo had started showing signs of aging back in 2014, when I notified Leader Dogs that I'd soon need a replacement guide dog. By early 2015, Leader Dogs had provided me with a smart young German shepherd named Sawyer. My schedule was far too busy for me to train with Sawyer at the Leader Dogs' campus in Michigan, so they delivered him right to our door.

The plan was for Romeo to retire and become our family pet, but Romeo had a tough time adjusting to his change in status. Each day, while Sawyer and I went to work, Romeo sat at home and moped as though he were clinically depressed. Like a lot of human retirees, he seemed lost without any work to do.

But Sawyer started giving us problems. When visitors came to the house, Sawyer would get excited and lose control of his bladder. It's called "excitement urination," and while most dogs grow out of it, Sawyer never did. Then, one day at our house, Sawyer lunged at a visiting three-year-old who had gotten in his face. That was the final straw. I couldn't have a dog that was capable of being so unpredictably aggressive. Sawyer had to go back to Leader Dogs. We didn't miss the peeing at all.

Romeo was very happy to put on the harness and go back to work. He was a little blind, and a little slow, but truly a reliable partner, at least for the time being. His vision continued to fade. One day at the airport, I could tell he was confused, which threw me off, because I couldn't be sure if he was taking us in the direction of the airport train, or if he was following someone with a bacon biscuit. As a team, Romeo and I were literally the blind leading the blind.

In March 2016, I received a young German shepherd guide dog named Bailey from The Seeing Eye, the world's oldest guide dog training center, located in New Jersey. I put Romeo back into retirement and started taking Bailey with me to work. About a week later, Romeo suddenly lost his appetite. We knew something was seriously wrong when he refused to even approach his food bowl—Romeo had never refused a good meal.

I asked Evie to take Romeo to the vet while I worked from home that day. Several hours later, I got the word from the vet that Romeo was fatally ill from a large tumor in his belly. I asked if we could bring Juliana by after school, so she could see Romeo one last time. The vet said Romeo was so sick that he might not last that long.

Evie and I took Juliana out of first grade in the middle of the day so she could say goodbye to Romeo. I brought Romeo one last Milk Bone, and although he hadn't touched his breakfast that day, he managed to enjoy this one final treat. Soon after, we said our goodbyes.

Our hearts were broken, and we cried for weeks afterward. To this day, we still talk about Romeo, who was Juliana's first dog and my trusted guide and companion for ten full years.

What we'll never get over is the timing of his passing, less than a week after Bailey arrived. It was as though Romeo had been quietly soldiering on because he knew that I needed him. He had hidden his illness from us all until he'd passed on his guide dog duties to Bailey. Then, at last, Romeo could finally let go and rest.

JACKSON IN THE HOUSE

Jackson Philip Foster joined our family on June 16, 2016. Like his sister seven years earlier, Jackson was delivered by caesarean section, but his birth was otherwise drama-free. At five pounds, he was perfectly healthy and ready to go home with us once Evie had recovered from surgery. We had been planning a second child for a few years, but had some problems conceiving. When we consulted with a fertility clinic, DNA testing revealed to us the cause of my eye condition, retinitis pigmentosa, or RP. Both my parents happened to be asymptomatic carriers of the same rare mutation of a particular autosomal recessive LRAT gene. Because I got a copy of this gene from each of them, I had only a 25 percent chance of being symptomatic for RP, which explains why my brother's eyes are normal. He came out on one side of the roll of the dice, and I came out on the other. As we expected, Evie's tests showed she doesn't have this genetic mutation, so none of our children are in danger of developing RP. But both are carriers, and before they have children of their own, their spouses should have their DNA tested.

We never appreciated what an easy baby Juliana had been until Jackson came along. Once Juliana had learned to crawl along the floor on her own, we clipped a bell to her so I could hear her moving around the house. We put the same bell on Jackson when he started crawling, and to me it sounded like one of Santa's reindeer with sleigh bells was running wild through the house. Jackson was everywhere and into everything. Only two days after he first learned to walk, we caught him scaling the outside of our stairwell like Spider-Man.

Juliana is a good sister who helps us keep an eye on Jackson, but he doesn't make it easy. Jackson has inherited my taste for mischief and daring, and as soon as he could talk, it seemed like he was talking his way out of trouble. He's very much like I was at his age, but with perfect vision. I like to think he's a better version of me—Chad 2.0.

I can't imagine what it's like to have a blind parent. I know that Juliana struggled with it when she was younger. It was very cute when she first learned to talk, and we taught her to say "Beep-beep, Daddy" whenever she was near me, just so I'd avoid stumbling into her. But children want to be seen and told how they look by their parents, and that's something I couldn't give her. "Daddy, are you watching me with your heart?" she'd sometimes ask, and all I could do was assure her that I was.

We'd play games together like hide-and-seek, but Juliana told me she didn't like watching me painfully bang into objects around the house while we were playing. Bumps and bruises are a part of blind life for me, but it was unsettling for Juliana to see. At night she would pray: "Dear God, my daddy's eyes are broken. Please send doctors to fix my daddy's eyes."

It frustrates me that I can't pass on to my children all the things that I know how to do but am no longer capable of doing. Sports were such a huge part of my early life that I'd love nothing more than to kick a soccer ball with my children or teach them to shoot a basketball. But when Juliana was eleven, I took her out of school for a week so we could go skiing with Paul and two of his daughters,

Anna and Emma, in Park City, Utah. By the end of the week, Juliana had graduated to the expert-level Black Diamond slopes. I couldn't have been prouder when, after she had trouble on her first try on a Black Diamond, she said she couldn't wait to get back on the chairlift and try it again. I'd like to think that she learned lessons on the slopes about grit, determination, and overcoming her fears that were far more important than anything she would have learned that week from her fifth-grade schoolwork.

More than anything, I believe children need consistency and follow-through on everything, even the smallest things. They need to understand that, as parents, our word to them is bond. If we say we'll do something, we must do it, and with such consistency that they know what to expect from us.

Evie hates to hear me say this, but I've learned this principle by training my guide dogs. I'm not saying that children should be trained like animals. Instead, I'm acknowledging that children, like all mammals, grow anxious and fearful when their environment doesn't feel secure and predictable. Any psychologist will tell you that when children are raised by unreliable parents who constantly send them mixed messages, those children often become adults with confused expectations about the world and about themselves.

That's not what we want for Juliana and Jackson. My overriding philosophy about child-rearing is to "expect greatness" from them, for the simple reason that if we expect mediocrity, that's likely what we'll get. If we hold our children to high standards and help them believe they are able to do something great, we raise the odds they'll lead more fulfilling lives.

I'd never expect either of them to follow in my footsteps and go into business. Their future career choices are not what matters. What matters most is the kind of standards they hold themselves to. Whatever they do, they should do it with an expectation of greatness. Have confidence, be bold, aim high, and do the work. Set lofty goals and then work toward them relentlessly. There is so much we can't control in life, but the one thing we all can control is our effort.

I've set a certain standard for my children's behavior that may be hard for them to meet. As they approach their high school years, they'll look at the obstacles I've had to overcome and find it's hard to excuse themselves for misbehaving or slacking off. I am naturally a strict parent, partly because my own upbringing was very strict, but also because, without my self-discipline, I'd be nowhere.

My dad was always a "tough love" kind of father. He tried to prepare me for life by always reminding me that no one cares whether I'm blind, and that no one was ever going to feel sorry for me.

My father also occasionally used a belt on me to correct my misbehavior. Although I'm not inclined to discipline my own children like that, my dad wasn't wrong about our uncaring world. When you bring children into your life, you know that the world can, at times, be an unkind place. They will struggle and feel pain, and that's not entirely a bad thing, because without struggle and pain none of us can grow. I'm concerned, frankly, when I hear about kids who seem to believe they have a right to feel comfortable in life. The uncaring world is bound to throw some misfortunes their way, and I fear they'll lack the resilience needed to respond.

At the same time, I recognize the need to constantly push for my own personal emotional growth. This is a lesson I learned during a leadership training exercise several years ago, when it suddenly struck me that I had a tendency to expect others to have my level of determination, and how unfair that was. I caught myself using my own experiences as a yardstick for how well other people were adapting to change in SRA. It bugged me that some got frustrated and wanted to give up rather quickly, and I was neglecting how my own hardheadedness and determination were rooted in a set of very specific and unusual circumstances.

Today I'm mindful that I need to extend this same respect to my children. Their upbringing is very different from mine, and that's a good thing. Will they grow up with the same level of grit and determination that I have? I don't know. What I do know is that I will need to meet them where they are as they grow up. So many things that I've had to scrap and fight for will come to my children very easily, and I will have to resist any urge to judge them for it.

The greatest gift I can give Juliana and Jackson is the gift of confident resilience. I want to coach them in attempting to conquer supposedly "impossible" achievements—first by starting small to avoid getting discouraged, and then by expanding their capabilities as they pursue ever larger goals. That's how our brains learn resilience. I can't teach Juliana and Jackson how to shoot hoops, but I can teach them the hard-fought lessons I learned myself after I lost my sight.

And when they fall down—which they will—they can count on Evie and me to help them pick themselves back up. I believe firmly in the need for children to fall down, but I also believe that the parents need to be there when they do, to teach them how to get back in the game. That's how they grow their confidence. That's how they learn to be resilient. That's how they discover they can turn any dark situation back toward the light.

RED HAT

Back in 2015, when SRA so generously sent me off to Harvard, I had no idea that the SRA culture I had thrived in and cherished so dearly was about to come to an end.

Around August of that year, my old employer CSC announced plans to merge its government services unit with SRA to create CSRA, a giant new company with $5.5 billion in annual revenues.

As the merger went into effect in late 2015, it started looking like it was really a CSC takeover shrouded in toxic office politics. All through the merger process, the CSC brass publicly stated how much they valued SRA's people, our culture, and our business development engine. Now that the merger was complete, many of SRA's teams were taken over by CSC folks and many of us were instructed to start doing things the CSC way. The SRA culture of empowering people to do great work fell by the wayside. In its place was CSC's cultural bias toward process and top-down control.

The irony was that CSC had targeted us because they envied our best-in-class growth and profitability. Now they were undermining the smart, idea-driven culture responsible for our success. It was like buying a goose that lays golden eggs—and then serving it up for Thanksgiving dinner.

I almost immediately came into conflict with my new supervisor, who was a very by-the-books kind of guy—the polar opposite of Ben and Joe, my previous bosses at SRA. My new boss told me I was too involved in all the different aspects of the bid, and that I should look at the pricing guidance clinically. He said I should only read what's been requested and avoid trying to shape the procurement. I should just set the bid strategy as the bid is written.

Every one of these directives ran against what we'd done to succeed at SRA over the previous seven years. I'd been involved in $45 billion worth of successful bids. We won those bids by shaping the bid proposal in our favor at every turn, every chance we got. I especially liked to influence the proposal in our favor, including the question-and-answer portion of the pre-solicitation period. I'd even launched a Financial Intimacy Maturity Model that brought data and analytics into presales conversations with customers. The objective was twofold. The model helped our customers structure the solicitation to better meet their objectives, and it also gave us an inside track on fine-tuning our pricing solutions.

Yes, my approach was more complex. It took in many more factors and relied on my own financial models, based on past successful deals, to help us make decisions. We did things the hard way at SRA because that's how you win. It's why we'd been able to kick off cash at an unprecedented clip in a shrinking federal IT market. But to pull this off, SRA supervisors had to trust their pricing people the way Ben and Joe trusted me and my teammates. The CSC method wasn't about trust at all. It was all about process and control.

I had three options, as I saw it. I could do it my boss's way (which would fail), secretly do it my own way, or fight to preserve the SRA way. None of those options were appealing to me. SRA was dead. My new employer, CSRA, was going backward, and I wanted no part of it.

It had been more than seven years since I'd done any job-hunting. Every week or so I'd get notices from LinkedIn about job opportunities, and I generally ignored them. Even when a recruiter from Google contacted me about a pricing job at the search giant, I said I wasn't interested. I was perfectly content at SRA. They'd given me raises, promotions, and a tremendous level of autonomy. They sent me to Harvard at their expense, and that experience changed my life. I had confidence in my leadership team and, most important, I knew my direct managers had my back.

So, when I started conversations with Red Hat, the big open-source software company, it felt to me like I was cheating on SRA. I had to remind myself that SRA was gone, and that my new employer was headed in a direction I didn't want to go.

I knew Red Hat would be a good fit when I asked the recruiter during a phone interview what specific role I was being interviewed for. "You know, we're not sure," she said. "We're just trying to hire the best talent we can find. Once we get you here, we'll figure out what role to put you in."

That kind of thinking—best people and best ideas first—really resonated with me. It was the exact opposite of the mentality I'd encountered with CSRA. A week earlier, someone in CSRA management had explained to me why certain employees from the SRA side of the merger would have to leave because there was no room for them on the organizational chart. I couldn't believe CSRA would let an organizational chart determine its future. There were some real SRA superstars that CSRA did not have a spot for, and they were leaving to join our competitors. Now we'd be competing for bids against some of our own best people, all because CSRA lacked the imagination to think outside the org chart.

Red Hat's business model impressed me: crowdsourced software development for enterprise-level customers. Red Hat doesn't have tens of thousands of software engineers on the payroll writing proprietary code. Instead, Red Hat relies on tens of millions of engineers around the world writing open-source code, which Red Hat refines and supports for its customers. Not only does Red Hat innovate at a faster pace than its competitors, it makes the world a smarter place by giving away the ideas to everyone for free. It was an inspiring mission. A purpose-driven company chasing a cause and not just profits.

I interviewed on the phone with Red Hat for a few rounds, and then during one interview, I was asked, "What was your biggest challenge in college?"

I paused for a second. *Do I have to be truthful and tell them I'm blind?* I decided I did, and I told them about my difficult six years at UT. For weeks afterward, I worried that I might be penalized for being honest. On the other hand, I was putting Red Hat to the test. Was this company mature enough to see past my blindness?

I made it to the next round of interviews and was invited to an on-site visit at Red Hat in Raleigh. Before going, I decided I'd prepare them in advance for any concerns they might have about my condition. That way we wouldn't be wasting the first fifteen minutes of every meeting talking about my guide dog and JAWS.

I thought long and hard about sending them the link to the video of my graduation talk at Harvard. I consulted with the recruiter, and together we agreed it was better to send it than not to. And that turned out to be the right move. The people I interviewed with loved it, and one of them called it a real differentiator for me. Ten years earlier, I don't think I would've had the self-acceptance and confidence in my vulnerability to send it.

Once I got my formal job offer from Red Hat, the former SRA people at CSRA started courting me to stay. Senior executives, including some at the very top, were asking me to reconsider. I felt torn. Emotionally, it was painful. I had grown to cherish many of the friendships I had built at SRA. I cared about many of the people there, and some of them I loved. We'd been through so much, building something that we were all proud of.

I asked Ben for his advice, and he came through for me, as he always did.

He asked me, "You're not having second thoughts, are you?"

"Honestly, it's tough," I said. "I'd love to stay here and have it work out."

Ben told me what I needed to hear. "Dude, as a good corporate citizen, I'd love to see you stay here," he said. "But as your friend, you have to take the job at Red Hat. It's a great opportunity. Besides, if it doesn't work out, I'm sure CSRA would love to have you back."

Ben always cared for me as a person more than he cared about a company or its profit margins. He helped keep me grounded about things that he knew were best for me and true to my spirit. It's why I—and anyone else from Ben's team—would gladly run through walls for him.

When I gave my notice, I told CSRA not to bother making a counteroffer. At the time, we were going through the question-and-answer period on a $150 million RFP (request for proposal) that would be my final bid at CSRA. We all like to go out on a high note if we can, and with this bid I made sure I had it all buttoned up before I left.

I ignored my supervisor's directions to stay in my lane. Instead, I did what I'd always done to win deals. I shaped the solicitation during the question-and-answer period and handed the proposal team a turnkey solution that included everything they would need to be successful. They knew the strategy, the pricing, the costs, the number of people, and the justification for the number of people we needed by task area. When the engineers came to me for advice, I handed them what they needed to write our tech solution.

Several months later, long after I'd left for Red Hat, I learned that CSRA had won that $150 million proposal.

As a win, it wasn't the biggest I'd ever had. Not even close. But it remains to this day one of the sweetest.

THE GIFT IN UGLY WRAPPING PAPER

Sometimes I like to say that blindness was a gift that came disguised in ugly wrapping paper. To me, blindness didn't seem like a gift at all until I'd unwrapped it and experienced how its trials and sorrows had made me a better person.

When changes came to SRA in late 2015, I can hardly express how painful it was to lose my happy home there, to say goodbye to brilliant and talented friends forced to seek jobs with our competitors, to have outsiders come in and vandalize what we all had built together with such pride.

With time, however, all this personal anguish proved to be the ugly wrapping paper concealing a tremendously valuable gift. Once I had processed my feelings of grief and loss, it was obvious that moving on from my old job was the best thing that could have happened to me.

As much as I loved SRA, it had been years since my role there had challenged me to stretch and work at the very edge of my abilities. Yes, every deal was different, a fun puzzle to be solved, but I'd already overcome most of the largest challenges in that job years earlier. Mastery can congeal into rote practice if you never need to raise your game. That's exactly where I was with my final winning deal at CSRA. It was all hard work, with few surprises or interesting new insights.

Red Hat, by contrast, was a constant source of surprises and interesting insights. Soon after starting there, I discovered that the organization wasn't at all ready for what I had to offer when it came to bidding big contracts.

Is that a disaster? Or an exciting opportunity? You can guess which one.

True to the ethos of crowdsourcing, Red Hat's culture invites every group to develop its own way of operating, without a single authority to provide oversight. The company naturally attracts a lot of independent thinkers, and its decentralized structure reflects that.

That works great for day-to-day collaboration, but when it comes to presenting our services in a unified way that's attractive to customers, getting lots of brilliant, independent-minded, and passionate people to align around a common approach is challenging. To make that happen, using influence is key. All of my experience convincing company executives to adopt aggressive pricing strategies came in handy when attempting to persuade others, which is crucial at Red Hat. And when we do align, we have so much momentum that we're an unstoppable force.

My first major project in the fall of 2016 was a pending bid with a huge telecommunications company that was one of Red Hat's most important customers. I saw an instant need to inject a sense of urgency into the team, because we were being entirely too casual about customer deadlines. There was a general complacency I was unaccustomed to, and it was clear to me we needed to sharpen our customer focus, or we risked losing this deal. It was a real scramble to fulfill the customer's requirements and get the master services agreement negotiated. We got it done in the space of three to six months, and ended up booking a nine-figure deal, the largest in company history at that time.

When I started at Red Hat, I did a lot of traveling for get-to-know-you meetings, and that took a toll on my guide dog Bailey, the successor to Romeo. In crowded places like airports, Bailey started showing signs of an anxiety disorder—panting and whining uncontrollably—and that was a problem. By June 2016, I'd contacted The Seeing Eye to see if I could find yet another replacement.

I explained to the trainer that I have no daily routine, and that Bailey might be fine for someone who did. Instead, I needed a dog that can fly overnight on long-haul flights to Asia, sit still during twelve-hour meetings, handle conferences and trade shows with tens of thousands of people, and be okay hustling up and down escalators on the way to catch planes, trains, and cabs, while visiting new and different cities all the time.

The trainer listened patiently to my list of particulars. "So you basically need a dog with a cape?" he asked.

I admitted it was a tall order, and he promised to keep an eye out for a truly special German shepherd. I had to wait nearly eighteen months before The Seeing Eye invited me up to New Jersey to meet Sarge, my super dog, in October 2017.

Within a month of Sarge's arrival, I knew he was the perfect guide dog. He's cool as a cucumber in any situation, and very rarely needs to be corrected. He's eager to do whatever I ask and hardly ever messes up. I get the feeling from him that he honestly just wants to make me happy. He's an amazing dog, my companion and guide to this day.

By this time, I'd given my keynote speech at least two dozen times to groups in the United States and abroad. My Harvard classmate Tomer, whom you may recall from the previous chapter, heads up the Atlanta Opera. He told me he was so moved by my speech in Boston that day that he'd decided to commission an opera based on my life experience. That knocked me out. Here's someone who's a pro at creating emotionally moving moments for audiences, and he thought my story packed that same kind of power. At this writing, the opera is in development under the working title *Sensorium Ex*, well on its way to a 2022 world debut.

I always meet a lot of interesting new people when I travel, and when they ask what I do, I'd gotten into the habit of telling them I'm "a corporate guy" and give the name of my employer. Now, for the first time, I wondered if I should also identify myself as a keynote speaker? One day, while on one of my many flights for Red Hat, I tried it out with my seatmate.

The woman responded enthusiastically: "You know, I can see that in you! You're so well spoken and articulate."

Getting such encouraging validation from a complete stranger felt so good that I began adding the keynote speaker line every time I met someone new. And a funny thing happened. As I told more people that I was also a keynote speaker, I became more comfortable saying it. I found that by sharing that part of myself right up front, I had some amazing conversations that wouldn't have happened otherwise.

One day at Red Hat, company CEO Jim Whitehurst came in to talk to the team. After the meeting, I approached him to ask his advice about writing and publishing a book. I'd enjoyed reading his book, *The Open Organization*, and I told him I really didn't know how to get started writing a book of my own. He referred me to his chief of staff, Emily Stancil Martinez, who put me in touch with Jim's literary agent. That's how this book came to be. Before I could even take the first practical step toward writing it, I had to see myself as a speaker and an author. I came to regret not doing this years earlier. How many more people could I have reached, how many experiences had I missed out on, by not opening my mind and my heart to my new identity? For years I'd felt uncomfortable when people told me I'd inspired them. But now that I'd accepted this role of being inspiring as part of my self-identity, that discomfort was gone. In its place was a new, wide-open space to grow. My thoughts, my words, my choices—even my clothes and my network of contacts—all evolved in the direction of this new, richer concept of my self, which I'd walled off for so long.

I've thought long and hard about the power of identity ever since. It all goes back to the stories we tell ourselves about ourselves. None of us are just one thing, though we tend to identify ourselves to others that way. When we do that, I think we risk boxing ourselves in, within the confines of our own minds.

Imagine how much personal growth you might experience if you told people a better story. Imagine declaring that you are also that one other thing you're afraid to say you are, that thing you imagine for your future self, outside your normal everyday role. When I learned to identify new dimensions of my future self, and began telling myself stories about that new self-identity, it unlocked possibilities that I'd been smothering for years. Now I find myself wondering from time to time, what else am I missing? What other parts of my future self am I overlooking?

After the big telecom deal was done, I took over the job Red Hat had hired me to tackle. I assumed the leadership role for the Global Deal Desk team. During one of the first few projects there, we realized that the company's entire pricing and discounting framework needed an overhaul. Its focus needed to shift from a transaction orientation to a customer-value orientation. Prices and discounting needed to reflect the value of the customer, not the value of the individual order.

The secret to a discipline of pricing is that if you use it to bolster your efficiencies in ways that benefit the customer, you can improve customer satisfaction and your profit margins at the same time. Communications between Red Hat systems and our customers are controlled by a series of request-approval-provision cycles. If we can cut out the need for half of the approvals in a given system, we can save money and increase our speed of getting deals done with customers, while making it easier for customers to add new services. Our job at Red Hat, and any other business, for that matter, is to make it easy for customers to spend their money with us. The easier it is for them to shop with us, the more they'll spend. It's not rocket science, but the devil's in the details.

The trouble is that initiatives of this sort usually come from Finance or Business Unit teams, not from the Deal Desk team. So, even though it made a lot of sense, it took considerable effort to sell it as a good thing for everyone—and not a power grab by that new blind guy from Atlanta.

I spent almost nine months traveling all over to get everyone on board. It was kind of like running a political campaign, but without the kissing babies part. We wanted to let folks know where we were coming from. We were not out to make their life harder. We just wanted to help.

There is no substitute for being physically present as you assure someone that your intentions are pure. Bridging that divide—between intention and perception—was everything. I had to explain how we took it upon ourselves to fix our pricing framework because we were looking at Red Hat through the eyes of customers and saw issues that others had been blind to. It had to be adjusted in a way that reflects current market conditions and customer needs, because it's not enough to be customer-focused these days. You must be customer-obsessed, or someone else will take your business away from you.

I had to make several trips to Boston with my team to meet with someone I'll call John. He had a reputation for being protective of his business area, and in this case, we completely understood if his guard was up. Historically, pricing and discounting changes of this kind were handled on John's side of the house, not ours. I'm sure there were those on John's team who thought, "Who do these guys think they are, trying to do our job for us?"

We started discussing these issues over the phone with John and learned very quickly that we'd never build rapport that way. If we could get John on board with us, we knew there were others in the company who would fall in behind him. So, we scheduled some flights to spend time with him in person, listening to his concerns and making sure he felt understood. As with most Red Hat units, his team's budget each year would be tied to our decisions in this new pricing and discount framework. He had plenty of reasons to be nervous.

I had to let John know that we were not trying to give away the candy store to customers. Sure, we wanted to be easier to do business with, but this wasn't going to be at the expense of our profit margins. It was understandable that John and his team wanted to poke holes in our recommendations. That's why we were there.

All I did at Red Hat that first year was stretch, stretch, stretch. It was a much better year than the one I would have had at SRA if nothing there had ever changed. The story I tell myself about this year of transition is that nothing in life is permanent. Time marches on, life moves on, and you have to move on with it. We can never be sure of everything when we're making our choices, and there was no guarantee that Red Hat would work out for me. But if you have a choice to make, I'd advise going for the one that creates the most opportunities for growth.

One such opportunity could be facing you right now. How can you be sure you're not staring at a gift disguised as an obstacle but you're too distracted by the crummy wrapping paper to notice? There could be a beautiful gift sitting right in front of you, if only you'd open your mind and your heart to receive it.

EIGHT

WE ALL HAVE BLIND SPOTS

"THAT DOG. IS HE COMING WITH YOU?"

The Uber driver's voice sounded muffled. I could tell he'd barely cracked his window open to talk to me.

"Yes," I shouted. "This is my guide dog. I'm blind."

"No, no," I heard the driver say. "No dogs. Allergies. I have allergies."

What followed was three minutes of utter futility, attempting to break down the driver's wall of ignorance. I always start off nice. I explained to him that the dog is clean and well groomed, but the answer was still no. Then I invoked the law. Discrimination against the disabled is a federal offense. He didn't care. Next came the threat.

"My flight is leaving soon," I said firmly. "If I miss it, I'm going to make you and Uber very sorry."

At that point, I heard the window roll shut as the vehicle drove off, leaving me stranded at the curb, with less than an hour until my flight. I contacted Uber to complain. They sent another driver right away, and I somehow made my flight that day.

When Uber was first introduced in the D.C. area around 2012, I was a big fan because it seemed like the ideal ride service for the blind. For me, it meant the end of fumbling with receipts or credit card machines in the back of taxicabs. Everything was handled through a smartphone app in advance so I could make a clean exit when the ride was over.

But Uber had a blind spot when it came to blind people with guide dogs. The company never trained its drivers to obey federal disability laws, and it refused to discipline drivers who broke those laws. As a result, I'd say about 30 percent of my Uber drivers turned me down once they saw me and my guide dog. Eventually, I joined a class action lawsuit led by the National Federation of the Blind. It took three years for Uber to arrive at a settlement agreement to train its drivers to accept blind passengers and deactivate any driver who discriminates against us. Becoming blind has taught me a profound lesson about the sting of discrimination that I don't think I would have learned any other way. As a privileged white man, I'd never known what it's like to be the object of prejudice until I was turned away by so many of the job recruiters visiting UT during my senior year of college. To be different is to be devalued time and again by uninformed people. I've had people tell me I don't look blind and ask who feeds my dog for me. I've sat next to Evie and listened to people talk to her about me in the third person as though I'm deaf or somehow incapable of speaking for myself.

When I traveled to Asia on Red Hat business, I learned what it's like to be different without the legal protections afforded me in North America and Europe. The discrimination against me and Sarge was extreme. In China and Singapore, in particular, I had to travel everywhere with a two-inch-thick sheaf of papers confirming that I'm blind and that Sarge has passed all the required health exams. Scheduling business lunches and dinners was almost impossible because so few restaurants would admit my guide dog, so we ended up eating outdoors a lot. But most onerously, I needed to complete an advance itinerary of my activities in the country, detailing where I'd be staying and where I'd be visiting, hour by hour. I don't understand any practical reason for this exercise, other than to discourage blind people from bringing guide dogs into those countries.

I've known for a long time that crude and false assumptions about who I am and what I'm capable of would limit my opportunities if I allowed them to. My father was right all along, that the world doesn't care. If I'm to make my way, I need to turn my disadvantages into advantages.

In fact, the author of the Americans with Disabilities Act, former congressman Tony Coelho of California, did just that. He never wanted to be in politics. His original career ambition was to be a Roman Catholic priest. But Tony suffered from epilepsy. During his physical for admission to the seminary, the doctor told him he was disqualified because epilepsy at that time was a sign he was "possessed by the devil." Tony subsequently turned his disadvantage into an advantage. He served five terms in Congress and championed the law that protects millions of his fellow disabled Americans from the discrimination he suffered as a young man.

Perhaps what I understand better than most people is that we all have blind spots, both literally and figuratively. There is an anatomical blind spot in every human eye at the point where the optic nerve attaches. We're unaware of this blind spot because our brains fill our field of vision with what we *expect* to see—things that are not really there. Our mental and emotional blind spots function the same way. They obscure realities that don't conform to our expectations and fill in the blank space with fictional assumptions. That's why it's so easy to devalue the contributions of others when they look different from us, think differently, or speak differently. We simply don't know what we cannot see when we cannot see it.

Our blind spots can blind us to our own opportunities in the same exact way. When we tell stories about ourselves that make us out to be victims of circumstance, we're not envisioning the full reality of the circumstances in front of us. Lots of us are dealt bad hands by life. The path from victim to visionary in any given situation can be a short one if you are willing to focus on what really matters and ignore your snap judgments—about others and about yourself.

As a blind person, I've had an advantage in this respect. I have been forced to reexamine all my assumptions and to learn how to see the reality of what is actually around me, stripped of preconceived notions. That same ability is available to all who are willing to recognize their own blindness, to explore every circumstance for its opportunities, and to treat every individual with respect and a genuinely open and curious mind.

DIVERSITY IS REALITY

Sometimes I think about the corporate recruiters who so rudely canceled my appointments with them the minute they met me at UT back in 2000. Have they learned better manners in the past twenty years? Have they at least been forced, like Uber, to behave better? Are any of them reading this and wondering what they might have learned that day if they'd treated me as a human being instead of a nuisance? Have any of them had their eyes opened to the inherent value that diversity brings to any human undertaking?

The recruiters that day judged me solely on my obvious weakness, that I cannot see. They had zero curiosity about my strengths. They didn't care how I might help lift their company up. All they could see were the ways in which I would slow them down.

On every team we join, there are things we're best at, things we're not, and some things we can't do at all. My own strengths in strategy and balancing competing priorities rely on other people on the team having complementary abilities for thinking outside our strategies or pointing out risks and downsides that I'm liable to overlook. We generate the overall best outcome when we respect one another's gifts and allow all team members to bring their talents to the table. Teams with a lot of diversity will ask more questions, probe different angles, and have a fuller view of the problem at hand. Everyone has different blind spots on a diverse team, helping ensure that potential weaknesses in the final product have been discovered and addressed.

A different dynamic takes place on teams when all the members are too much alike. Decisions may come quicker and more efficiently when everyone appears to be on the same wavelength, but the decisions are likely to be flawed because group members also share similar blind spots. This phenomenon, called *groupthink*, has been found at the roots of the 1986 *Challenger* space shuttle disaster, the 2001 Enron bankruptcy scandal, and the 2003 invasion of Iraq.

There's a farming term called *monoculture* that describes planting one kind of crop exclusively in a broad area, which allows much greater speed and efficiency in seeding and harvesting. The result is greater yields and bigger profits, but there are also risks. When you have a large area filled with just one kind of crop, disease can spread like wildfire, as it did in Ireland in the 1840s, when a potato blight caused the Great Famine that killed an estimated one million people.

Monocultural teams and organizations can suffer from the same effects when it comes to the spread of bad ideas. The chief reason teams lacking in diversity arrive at decisions so quickly is because members feel pressure to conform to consensus. A group comfort level forms around one of the team's options, and preserving that solidarity becomes valued more highly than fully exploring alternative options. Team members wind up withholding their contrary views for fear of spoiling the consensus.

By contrast, teams with diverse memberships put everyone a little outside their comfort zone, because that's where we all grow. Practicing diversity takes deliberate effort, without a doubt. Team members need to be humble enough to listen to opinions that differ from their own and recognize that no one person has all the answers. But the effort is worth it, because a true diversity of thought inspires fresh thinking from a much broader range of perspectives.

Research shows that productivity and creative outcomes soar when we build diverse teams around people's strengths. A McKinsey study found that the benefits of diversity show up directly in the bottom line. At US public companies, profits averaged nearly one percentage point higher for each additional 10 percent of racial and ethnic diversity on senior executive teams.

It's easy to confuse diversity of thought with just having a lot of people around the table who look different. I can't confuse the two because I'm blind. True diversity of thought is something I can detect in the meeting room, undistracted by how everyone looks and acts. My blindness has made me more deeply contemplative than I had ever been before, and I hold it as a gift. To the extent that I've been a visionary, it's because I cannot see.

In Malcolm Gladwell's bestseller *Blink*, he describes how orchestras have come to recognize their blind spots by embracing the gift of blindness during auditions. The book tells how orchestras had normally hired big men for the French horn because everyone assumed that playing the instrument takes great physical strength. But when the Munich Philharmonic held a series of "blind auditions" by having auditioners play their piece from behind a screen, the Philharmonic's leaders had no choice but to close their eyes and just listen. They unanimously decided that one auditioning French horn player was the best—and out from behind the screen stepped a petite young woman.

To enjoy the full benefits of diversity, we must exercise empathy and accept that people's beliefs reflect the accumulated experiences in their lives. We may not be able to fully understand their experiences, but we should consider that we might think alike if we'd had similar experiences. By now you've no doubt imagined who you'd be if you'd lost your eyesight at age twenty-one. You owe that same level of empathy and insight to everyone you care about. *Who would I be if I had experienced what she has?* You owe it to yourself to ask that kind of question about all your workplace team members, too—especially those with whom you may clash.

Diversity is the engine for innovation. Every industry on the planet is being upended by new ways of doing things. If we know that diversity of thought nurtures change, and that change drives innovation, we need to harness all the virtues and benefits of true diversity in order to thrive in the years ahead. It's key for all of us to give every person a chance to exercise their talents, to help them feel welcomed and valued to contribute. And as diverse people, we also need to help others get comfortable with us. We need to speak up for ourselves and make our needs known and our strengths known.

When we deliberately include diversity without unconscious bias, and thoughtfully engage others, we can inspire and empower each person while fueling the innovation our businesses need to succeed in a fast-changing landscape. How can each of us be more aware of our mindsets, language, and behaviors to ensure that each individual feels compelled to shine his or her brightest light on our most pressing challenges? Because, when we all bring our best, human potential benefits the most.

I feel fortunate to be working for Red Hat, where this global vision of diversity shapes our everyday reality. We rely on the pioneering open-source code created and tested by millions of software engineers around the world, all of whom care only about continuous innovation and superior products, regardless of who contributed which part.

It's a regular reminder that nothing truly groundbreaking is ever achieved alone. No matter how great your personal success, you need a team of people behind you, elevating you and your work. And the more diverse our teams are, the stronger, more resilient, and more durable our own success will prove to be.

YOUR VISION OF GREATNESS

One week before I was due to give the publisher this manuscript, my friend and mentor Ben Gieseman passed away at his home in Virginia after a two-and-a-half-year battle with cancer. He was only fifty-three. Ben left behind his loving wife, Beth, and three wonderful children, Josie, age fourteen, Max, eighteen, and Evie, twenty. His obituary noted that he was "a highly respected and much loved colleague" who "always enjoyed a laugh with his coworkers whom he considered friends." My first laugh with Ben appears in the opening of this book. It would prove to be the first of many.

I will miss Ben terribly, the man who put so much faith in me, who had my back so many times, who even offered me a place in his home during my time of need. Now that I've lost my friend forever, I can emerge from my grief with several stories to tell myself about Ben's legacy. How can I best remember Ben in a way that fully honors his contributions to my life?

Maybe Ben's final gift to me is a reminder that none of us are here forever, and that I have to keep up the sense of urgency to keep growing, to continue improving, and to use my blindness as a gift to others, to help others see. To see their own blind spots. To see their own possibilities. To see their gifts that are disguised in terrible wrapping paper. To see the obligation they have to open their gifts and share them with the world.

As I write these words, we are in the middle of a global pandemic, with no idea when or how life will return to normal. My family and I have been, like many, quarantined at home. Juliana has completed her fifth-grade classes online, Jackson is keeping us busier than ever while not in preschool, and I'm working long hours at my laptop and on the phone. Evie is taking care of all of us, and somehow we're able to balance being caregivers, teachers, parents, providers, and professionals—all with traditional work-life boundaries blurred by being at home 100 percent of the time.

As chaotic as it has been, we count ourselves lucky. No one in our close circle of friends and family has been medically affected by the virus. And thanks to the nature of my industry and the nature of my job, our income has been unaffected. We as a family are not suffering the way millions of other people are, all over the world. Many readers of this book will have lost loved ones in the pandemic. They will have lost their jobs, and some may have been evicted, or lost their homes and everything else they've worked for their whole lives. Or they've come down with COVID-19 symptoms, and they've been disabled by the ravages of the illness.

Is it too much to ask that we each try to see the pandemic as a gift in terrible wrapping paper? I hear politicians express hope that the world will come back stronger than ever, but that's cold comfort if you're sick from the virus, or lost someone close to you, or lost your home, your business, and all your economic security. To tell a better story about the pandemic will be a struggle for most of us, and for some of us, it will feel impossible. But it's necessary if you want to move ahead in your life. You have to count every small gift that's granted you.

Yes, the pandemic is a global disaster, but we must believe that it's also more than that, because that's the truth. Only you know the exact toll it's taken on your life—financially, emotionally, physi-cally. It is essential that you continue to grieve those losses. But you must also count the blessings of the months spent in lockdown and social distancing. Those blessings are real, even if they're dwarfed by your losses. For me personally, I am grateful for the time with my family that I don't otherwise get when I'm traveling for business so often. I'm thankful that the time I once spent in airports and hotels, I can now spend with loved ones.

As surely as I acknowledge the benefits of my blindness, we all need to acknowledge the many ways we've grown from the experience of 2020. We must take account of the new relationships we have formed in these difficult months, and all the tough times when we've really connected with our inner strength and resolve. That's how we can summon the resources to find a way to a better future for ourselves and those we love.

I've said many times before that one reason I feel such a thirst for life is because we are all here for a very short time. I say this not in a morbid way, but as a motivational nudge. None of us know how much time we have left. If I'm lucky enough to live a long life, I want to be able to look back and know that I tried all the things I aspired to. I will have failed at some, and succeeded at others, but I will know that I gave it my all and left nothing to chance.

The biggest blind spot that holds us all back is the difficulty we have in seeing ourselves clearly, stripped of delusion, self-deception, and unnecessary pride. It can be painful sometimes to see ourselves in this way, but until we do, we have no possibility for self-acceptance. And that's a trap. Lack of self-acceptance is inevitably the path of depression, defeat, and victimhood. It was the path I was on when I could not get beyond my anger and frustration over going blind.

Because my personal situation was so dire, I was forced to make a conscious effort to distinguish what I could influence about my circumstances and what I could not. I finally determined that my blindness was beyond what I called my sphere of influence. There was nothing I could do about it except tell myself a story of acceptance about it. The primary story was that I would not use my blindness as an excuse for not trying, because excuses are for losers, and I was not a loser.

Effort is always within my sphere of influence, so I would try with everything I had to get what I wanted. When I failed, I would not make excuses, I would just try again. And that's what I've been doing for a little more than twenty years, from the time I returned home to Knoxville with Miles by my side.

I had been a blind guy who felt like a victim. I had to adjust my self-image and tell myself new stories about all the important facts in my life that were either inside or outside my sphere of influence. With acceptance of the things on the outside, and with determination toward the things on the inside, I've developed a vision of greatness for myself. It's one that I have not stopped reaching for. It's how I've changed my mindset and life story from victim to visionary.

I cannot change how I was born, so I don't even think about it. There is so much to do inside my sphere of influence that I no longer fret about all the things on the outside. Instead, I work hard at holding myself accountable for what I can affect in pursuit of my vision of greatness.

Maybe you don't feel like you deserve a vision of greatness for yourself. Maybe you think I'm delusional to think that I have one. The truth is that it's within everyone's power to have a vision of their own greatness, their own best selves. And in our hearts, we all have a sense of who that person is inside of us. Whatever the facts of your life are, consider that with a slightly different story to tell about those facts, you might find a vision of yourself that you didn't think was possible. I remember myself in my dorm room at Leader Dogs for the Blind, the night before I was paired with Miles, and believe me, what has happened in my life since then was the furthest thing from my mind.

It's hard for me to describe the surge of energy and focus that seized me when I freed my mind of my negative thoughts about how I'd been victimized by blindness. When I was able to embrace my blindness—first as a fact, and then as a gift—it opened up all the mental space I needed to reimagine a better life for myself and all the people around me.

My gift of blindness has made me happier than before. I'm more intentional and more fulfilled than when I could see. I'm also less emotionally reactive. The gift is so powerful, I feel compelled to use it to help others see how much more there is to life when we tell ourselves the right stories. My gift has caused me so much pain, and through that pain it has awakened the awareness that whatever the circumstances we face, there is always a path forward, whether we can see it or not.

What is your vision of greatness for yourself? What are you going to do with your life to make a difference in the world? I believe everyone has something they can do to pay it forward. What are the talents and skills you bring that will lead you on the path to your own personal greatness? You only need to raise your awareness of them and start using them. You will find your way.

That's my closing wish for you and for everyone. I hope you unwrap your gifts and through them discover your purpose, your passion, and your personal vision of greatness. I hope you'll pursue that vision with joy, with generosity, and with blind ambition.

VICTIM TO VISIONARY RESILIENCE EXERCISES

LOSING MY EYESIGHT AT AGE twenty-one was such a crushing blow that I quickly fell victim to patterns of negative thinking and self-defeating behaviors. Who wouldn't fall prey to unproductive thoughts after going blind in their twenties? I could have remained a victim of circumstance for the rest of my life if I'd let those thoughts and behaviors persist. Instead, I am leading a life today that I would have considered far beyond my reach before I went blind.

I've spent years deliberately retraining my mind to break free from the self-destructive thoughts that were holding me hostage, and I've included these techniques on the following page to help you break free and move your mindset from victim to visionary. I would never say the path was easy, but these six techniques I've developed to help me are deceptively simple.

1. Choose happiness in your daily life.

2. Recognize when you're letting excuses hold you back.

3. Practice moving out of your comfort zone.

4. Change your perspective in the face of setbacks.

5. Set a Life Vision for yourself that inspires your daily choices.

6. Tackle each obstacle that stands in the way of fulfilling your Life Vision, one task at a time. Stay focused!

Life is lived one day at a time. If you apply these tools to both your problems and your opportunities on a regular basis, you will find that positive thinking and high productivity are habit-forming. As you put your advantages to work toward your vision of greatness, you'll also appreciate the unique perspectives you've gained from your disadvantages. Your gifts will reveal themselves, no matter how ugly their wrapping paper.

You'll want to revisit some of these exercises every day. Others you'll want to turn to in times of trouble, when you're facing a difficult dilemma or a big challenge. These offer a structured process to open up your mind and find out what's in your heart.

These exercises will help you cultivate a more resilient mindset, so you are more capable of accepting what you can't change to adapt and overcome whatever life throws your way. There's no magic fairy dust in these exercises. You get out of them what you put into them. Be candid with yourself. Be kind to yourself, but be assertive, and most importantly, be relentless.

HAPPINESS
IS NOT A FEELING

(CHAPTERS 1 & 2)

YOUR CIRCUMSTANCE

Negative thoughts and feelings of
frustration and sadness.

YOUR OBJECTIVE

Tell yourself better stories about your circumstances.

APPENDIX

THE FIRST FEW YEARS AFTER I went blind, I struggled to accept my condition. I discovered that I had never envisioned or aspired to be blind, so my high hopes and self-identity were devastated when I lost my eyesight. It was during that time that I learned how happiness is anchored to our perspective, not our facts. It's a choice we make. We tell ourselves stories about our situation, and we can be deliberate about choosing stories that make us happy and power us forward, or opt for stories that hold us back.

Once I was able to accept the reality of my blindness, I could also take a factual inventory of all my skills, talents, and other advantages for which I could be grateful. I had my physical health and strength, a talent for analyzing financial information, and a supportive family, among many other gifts.

These advantages are just as true as my problems, and they cast my problems in a new light, which gave me the tools to choose stories that supported my happiness on a daily basis.

1. Write out a description of what's making you feel frustrated and hopeless in any of these four areas of your life. Work, relationships, personal growth, health/fitness.

2. Write out the story you tell yourself about each situation.

3. For each situation, write out the worst-case scenario. How much worse could it be?

4. Now write out all the other abilities and advantages you could be grateful for in that area where the situation is causing you pain.

5. For each situation, write down the control you have in the situation no matter how minor. For me, it was as simple as being able to learn how to use a computer despite my blindness.

6. Compare the list of gratitude items against your feelings of frustration and hopelessness. Write new stories about your situations that are hopeful and empowering, that account for unchangeable facts.

7. Start a gratitude journal where each night you write down the top three things you're grateful for every day. We do this at home to help our children develop the muscle memory of gratitude and perspective.

Take your exercise worksheets with you anywhere on the go. Visit ChadEFoster.com/exercises to download and save your worksheets on your preferred device, or just print them!

EXCUSES ARE FOR LOSERS

(CHAPTER 3)

YOUR CIRCUMSTANCE

Anxiety and feeling defeated by difficulties.

YOUR OBJECTIVE

Distinguish what is within your ability to change.

I FELT BRUISED AND BATTERED after I lost my first job out of college. At that point in my life, I had plenty of excuses for giving up. I could have consoled myself with a story about how the workplace is unfair to people with disabilities, which is absolutely true. But that's not the story that would propel me forward.

To be a visionary and not a victim is to envision with clarity the difference between what can be changed and what cannot. You need to accept the aspects of your situation that are beyond your control without judging yourself. Then you must identify what actions are within what I call your "sphere of influence." Once you've told yourself new empowering stories about how to change the aspects of your situation within your sphere of influence, you can also tell yourself stories of acceptance about what you can't change.

As Dwayne "The Rock" Johnson says, "Success at anything will always come down to this: focus and effort, and we control both."

1. List the top three or four situations in your life that are a source of anxiety and difficulties.

2. List the plain facts at the source of each of these situations.

3. Which ones of them are outside of your sphere of influence? What would it take to make these circumstances look good, so it's easier to accept them? Be bold here. Paint a vision that inspires you.

4. Now note which of those facts are *inside* your sphere of influence, allowing you to take action to improve

the situation and drive toward your inspiring vision painted in step 3.

5. For each situation, list the top two goals. Use positive and not negative action-oriented goals. For example, if your goal is to lose weight, instead of listing "lose weight," list "eat a set number of calories each day" and "exercise for a set number of minutes per week." The more specific the better. Consider SMART goals that are: Specific, Measurable, Achievable, Relevant, and Time-based.

6. Now, write down the step-by-step tasks within your sphere of influence that will help you reach each goal. Action is key. Progress prevention is caused when we have good intentions with no actions. As Tony Robbins said, "A real decision is measured by the fact that you've taken a new action. If there's no action, you haven't truly decided."

7. Start a thought journal where each day you write down the thoughts you're having when the distressing situations listed above arise. This simple practice helps you pay attention to the thoughts and feelings underneath your anxiety.

Bonus: Regular mindful breathing is another way you can strengthen your awareness of inner thoughts and feelings. Just as routine exercise causes your physical conditioning to improve, the consistent practice of mindfulness causes your mental conditioning to improve.

It doesn't require hours every day. Only ten minutes of mindful breathing has demonstrated enhanced cognitive function—often leaving people happier, more emotionally balanced, better with concentration, and exhibiting an improved working memory.

In fact, brain imaging techniques have revealed that after only eight weeks of mindfulness, the amygdala (the brain's fight-or-flight center) appears to shrink, while the pre-frontal cortex (the part of the brain associated with higher order functions such as awareness, focus, decision-making) grows thicker. Mindfulness redirects the activity in the brain from the primitive, reactionary part of the brain to the part of the brain responsible for executive functioning. This helps us cope with stress while boosting our awareness and focus.

Don't worry if you've never done it before—there are many great tools available for it. The one I use is Headspace (www.headspace.com).

LIFE BEGINS OUTSIDE YOUR COMFORT ZONE

(CHAPTER 4)

YOUR CIRCUMSTANCE

You are facing new challenges in
unfamiliar situations.

YOUR OBJECTIVE

Overcome fear of risk and the unknown.

IT CAN BE TEMPTING TO seek a comfortable life that follows the path of least resistance, but that's not how we learn and grow. Comfort has never been an option for me. I spend every day outside my comfort zone, so I get more practice in the daily experience of learning and growing than most people.

What I've found is that the further I venture into my discomfort zone, I keep finding new challenges that open up exciting new possibilities I never would've learned about any other way. So try doing things that keep pushing you toward your new edge of discomfort. Regular physical exercise, by the way, is a great way to acclimate yourself to the value of pushing beyond your comfort zone. When you stress your body with daily exercise, it builds mental toughness and keeps you better prepared to handle the stresses that life sends us every day. Your body gets used to the feeling of stress from the physical exertion, so when your body feels the anxiety from life's natural stressors, you're better equipped to handle it.

1. Name three areas of your life where you've avoided doing something fun or interesting because it's outside your comfort zone.

2. For each area, identify a dream or goal that seems out of reach.

3. For each dream/goal, identify an action you can take tomorrow that gets you incrementally closer to reaching it. This is not about being bold or daring. It's about progress.

4. Try to do one thing every day for the sole reason that it's outside your comfort zone. Get comfortable with discomfort.

5. Start a confidence journal where each night you write down three things capturing how you displayed skill, effort, and/or determination.

Bonus: Start a habit of regular exercise. In the beginning it's not about how much you exert yourself.

In fact, that is counterproductive as you'll be too sore to return to the gym. Instead, start slowly. And, it doesn't even matter if you work out at all the first few times. Consistently showing up to the gym is a victory. If you change your lifestyle to include regularly going to the gym, eventually you'll start working out and get into shape.

And when you start seeing yourself as a mentally disciplined person who routinely challenges your physical limits, your self-image begins to reflect the discipline, mental toughness, and determination you display to make exercise a daily habit. You'll feel stronger, more confident, happier, and better prepared to handle whatever life throws your way.

WHO WANTS IT MORE? NOBODY!

(CHAPTER 5)

YOUR CIRCUMSTANCE

Facing big decisions with a lot at stake.

YOUR OBJECTIVE

Overcome fear of risk and the unknown.

DURING MY DARKEST DAYS, OR when I've found myself at a cross-roads, there's always been a big new chapter in my life ready to open up. A crisis in your life reveals what's most important to you. The critical factor is to take action. To develop a visionary mindset, you must test your fresh thinking in the real world and experience the results.

1. List the top three ambitious goals you think about most often. Be bold here.

2. Are you afraid of success? Are you comfortable with reaching these goals?

3. Or, are you hesitant about putting yourself out there and not quite making it?

4. Write down how you feel deep inside when envisioning the prospect of pursuing and reaching these goals.

5. What scares you more—failing in pursuit, or not knowing whether you could've made it? Which fear drives you?

DARE
TO BE GREAT

(CHAPTER 6)

YOUR CIRCUMSTANCE

Confusion over your goals in life.

YOUR OBJECTIVE

Having a Life Vision that will inspire your
achievement of personal greatness.

WHAT I DISCOVERED WITH MY speech at Harvard is that when you keep pushing yourself outside your comfort zone to overcome your fears of the unknown, your purpose in life can reveal itself in surprising ways. And when it does, you have to seize your personal vision of greatness and nurture its growth.

Blindness has not prevented me from pursuing one vision of greatness after another for the past twenty years. At first, all I wanted to do was excel at my job and serve as a role model for other blind people and the disabled community. With time, my vision expanded to include leading and teaching teams in the workplace. Now, my vision includes using my story and life lessons to help the greatest number of people—whether it's through this book or keynote presentations.

My fellow Knoxville native, the great film director Quentin Tarantino, says that the job of moviemaking is "explaining your vision." The same is true when directing your life story, and the stakes couldn't be higher. You need to understand your vision, be able to see it, describe it, explain it, and inspire others in pursuit of it.

Use these steps to paint a vision of greatness for yourself. Be bold. Dare to be great. Aim higher than you think is possible. You will be surprised at just how far you can go. Once you've painted that vision, you should believe it's as real as the water you drink. It should permeate into your vocabulary. The words you choose should reflect your vision. Those words will become your actions, and your actions will drive you to your outcomes, and to new visions of greatness that await you in years to come.

1. When you look back on your life many years from now, what are the three most important factors for

your "success"? What can you live with and live without?

2. When you look back on your life many years from now, what are the three most important factors to your happiness? What is really important in life?

3. What are the three things you should change now that currently take priority but didn't make your list above?

OBSTACLES EQUAL GROWTH

(CHAPTER 7)

YOUR CIRCUMSTANCE

Unforeseen difficulties in pursuing your goals.

YOUR OBJECTIVE

Tell new stories about the obstacles in your way.

APPENDIX

IF YOUR VISION OF GREATNESS is bold enough to inspire you, then the path in pursuit of your vision will inevitably be littered with obstacles. There's so much you don't know or can't do by yourself to get where you want to go.

The best story to tell yourself about obstacles is that each one represents a new opportunity: to learn new skills, to find valuable allies, to open up new possibilities you couldn't have seen any other way.

Take all the facts about your obstacles and perform a gap analysis to pinpoint what it will take to overcome each one. What will be your next step for tomorrow? Next month? Next year?

With each obstacle, tell yourself the story about your vision of greatness, that your vision is as real and solid as the ground you're standing on, that it's out there just waiting for you to fulfill its promise.

Believe it. Assume it. Demand it.

1. List the top three obstacles preventing you from being happy and successful.

2. For those obstacles outside of your sphere of influence, list the possible stories to explain those obstacles to yourself—both good and bad.

3. Now, decide if you want to live a life according to the good stories, or the bad stories. It's your choice.

AUTHOR'S NOTE

The names of some of the individuals featured throughout this book have been changed to protect their privacy.

ACKNOWLEDGMENTS

So many people have helped me through the years that it's impossible to thank them all by name. I will do my best to recall them all in these "Acknowledgments" and hope that I've hurt no one's feelings if I've unintentionally left them out.

First of all, thank you to Jim Whitehurst and Emily Stancil Martinez for connecting me with the team I needed to make this book possible. Without your generous support and advice, *Blind Ambition* might have remained nothing more than a dream in my head.

Thank you to Tony Coelho, former congressman from California, for being a great mentor to me personally and for being a trailblazing leader for the entire disabled community. Your championing of the Americans with Disabilities Act outlawed discrimination against us for the very first time in history. You opened the door for tens of millions of people to be treated with dignity in everyday life, to find fulfillment in work and pursue the American Dream.

Thank you to Bill George for your unique insights and teachings at Harvard, which inspired me to use my life's greatest obstacles to help others achieve higher outcomes.

Thank you to Gordon Gund, Jason Menzo, Benjamin R. Yerxa, and the entire team at the Foundation Fighting Blindness for your commitment to bringing breakthrough discoveries and treatments to market for blind people. Thank you to Gordon for your unique and invaluable philanthropic support.

ACKNOWLEDGMENTS

Thank you to Eric Damery at Vispero for your decades of dedication to JAWS for Windows and all the other assistive technologies that create new possibilities for visually impaired people worldwide.

There are far too many special people at Red Hat to mention here, so I will just single out five great Red Hatters who have been so helpful in getting me acclimated to the most astounding company culture on earth: Carl Trieloff, Franz Meyer, Gitesh Vohra, Jordan Childs, and Terry Tomlinson. All of you exemplify Red Hat's open-source culture, which combines bright minds with caring souls.

Thank you also to Red Hatters Arun Oberoi, Dirk-Peter "DP" van Leeuwen, Kirsten Kliphouse, Paul Smith, and Werner Knoblich for inviting me to speak to your Red Hat field teams during our annual sales kick-off conference.

The people at SRA have such a special place in my heart. I will always be deeply grateful for how you welcomed me in, had faith in my capabilities, and gave me the space to grow and learn.

Thank you to Ben Gieseman, who hired me at SRA, and whose passing in 2020 left me feeling bereft of a dear friend and trusted mentor. To the entire Gieseman family, please know that Ben's generous spirit and his unique mix of savvy and humility lives on in my memories of him.

Thank you to Joe Readyhough for challenging me and trusting me to drive the largest and most complex deal strategies in the company, as well as the mentorship and opportunity to grow. To Lu and Julie Pham, special thanks for all the good times during my time at SRA and for being there for me in my time of need.

Thank you to all the SRA executive team members for all their support and guidance: Adam Rossi, Bill Ballhaus, Dave Keffer, Dave Rue, David Cox, Deb Alderson, Donald Hirsch, George Batsakis, Gilbert Dussek, Jake Stenzler, Jeffrey Rydant, Jim Enicks, Jim McClave, John Anderson, John Levy, John Ludecke, John Luongo, John Patrick, Mark Connel, Mark Ginevan, Max Hall, Michele Engelhart, Pat Burke, Paul Nedzbala, Rebecca Miller, Scott Ayers, Steve Liptak, Steve Tolbert, Tim Atkin, Tim Day, Todd Morris, Tom Hutton, and Tom Nixon.

And thank you to all my friends and colleagues on the SRA growth teams for your good humor and team spirit through all the agonies and ecstasies in pulling needle-moving deals together: Allen Deitz, Allison

ACKNOWLEDGMENTS

Patrick, Andrea Hitcho, Bob Becker, Bob Hale, Bob Heckman, Brad Greenfield, Brian Cross, Brian Michl, Bryan Polk, Carl Willis-Ford, Chad Ehrmantraut, Cheryl Kaufman, Chris Albrycht, Chris Cecka, Chuck Crotty, Craig Cheney, Craig Wilson, Daryl Davis, David Mutryn, David Page, Dawn Colby, Doyle Choi, Emily Crespin, Eric Kurzhals, Gary Kerr, Gina Wolery, Gordon Levy, Jamil Abdel-Jalil, Jeanette Lucky, Jennifer Bowman, Jeremy Berens, Jessica Morgenstern, John Mays, Karen Hertel, Karma Temple, PhD, Kate Poppert, Kathleen Yoshida, Kelley Artz, Kevin Marshall, Lenny Zuriff, Majed Saadi, Mark Barnette, Mark Grant, Mark Mayhugh, Marki Louis, Matt Modica, Matt Parker, Michael Bartholf, Michael Goodrich, Michael Ryan, Miguel Gorordo, Mike Thomas, Morna Green, Nick Trzcinski, Patricia Gietl, Philip Bayer, Phillip Dicken, Priscilla Golden, Ray Good, Rebecca McHale, Renee Brown, Rob Blake, Robert Forster, Robert Smallwood, Russ Gilbertson, Sandi LaCroix, Sandra Ambrose, Sandy Gross, Sarah Coburn, Selina Hayes, Shawn Hammer, Shirley V. Louangamath, Sif Lazizi, Steve Hyde, Steve Shively, Subin Punnoose, Tammy Ollila, Teresita Maldonado, Terry Griffin, Thomas Olson, Tom Oliver, Tom Pugliese, Tony Meyer, Vinnie DeVito, Wendi Mitchell, Will Harkins, Will Watts, and Yvonne Costello.

At CSC, thank you to Ash Sharman, Bill Hutton, Mark Lees, Marty Napier, Richard Ricks, and Tom Bailey for all your support and mentoring during my days there. Thank you, Ash, for taking me under your wing and teaching me so much about the business of IT. Thank you, Bill, for sharing all your wisdom and your impeccable balanced judgment. Thank you, Mark, I learned a lot about customers while on your team. Thank you, Marty, for helping me grow into my new role on the team and for your superstar abilities to work with people through such complex transactions. Thank you, Richard, for welcoming me to the organization and for your ability to inspire others to exceed their own expectations. And thank you, Tom, for giving me a chance to showcase my skills. You are truly a special person, and I'm a better person for having gotten to know you.

At Bender Consulting Services, thank you to Carol Howard, Jim Homme, Joyce Bender, Mary Brougher, Mike Gravitt, Neal Lutz, Paula Ballog, Ricco Brusco, Wade Churchfield, and the whole Bender family for your inspiring leadership on behalf of the disabled community. I am proud to have been a part of it.

ACKNOWLEDGMENTS

At Bartimaeus Group, thank you to Don Olson, Doug Lee, Jon Avila, Glenn Smith, Mark Reumann, Mary Bullock, Tamas Babinszki, Victor and Karo Tsaran, and Yara Salazar for all your help and support in executing our special mission together.

Thank you to Nancy Goodwin for teaching me the ropes with thoughtfulness and good humor while at the Georgia Department of Labor.

Thank you to Kathryn DeBusk for all your help with support services at UT and to Suzan Murphy for giving me the added time and attention I needed in your class at UT. You balanced your empathy with high expectations for me during a crucial time in my young life, and it's made all the difference.

Thank you to Richard Moon and the Halls Community Lions Club for sponsoring me for Leader Dog training and to James Odom Jr., Rich Guzik, and the entire Leader Dogs organization. Special thanks to Rich for introducing me to Miles, my first guide dog and loyal companion for seven years. And thanks also to Jim Kessler at the Seeing Eye for finding me Sarge, my Super Dog, and to Glenn Hoagland and the entire Seeing Eye organization. To Steve Schnelle, thank you for the time we spent together at Leader Dogs and for patiently explaining to me that using a computer without eyesight is possible.

To Lisa Hughes, thank you for welcoming me to Harvard Business School with open arms and an open heart. Thanks to all my classmates at HBS for making me feel welcome and for electing me as your graduation speaker. Thank you to Garrison Wynn for giving me world-class coaching and advice on that speech.

To Sonia Marzec, thank you for being my first skiing coach, for putting up with my stubbornness, and for keeping me alive on the slopes. Thank you to Jim Lunay, whose early encouragement for us to go skiing in Aspen changed my life. Thanks also to Jeff Hauser at Challenge Aspen and Mark Fisette in Park City for all your ski guidance and instruction, and for showing me how the real barriers only exist in our minds. And thanks to Paul (and Shawna) Hibben, for our friendship, for being my ski guide and videographer on so many downhill runs, and for raising our game together on the Double Black Diamond slopes.

Thank you to Chuck Stidham, Adam and Kimberly Donovan, and Ken and Stephanie Price for your loving friendship over the many years.

ACKNOWLEDGMENTS

To Esmond Harmsworth, thank you for coaching me throughout the entire publishing journey, for keeping us on track when it seemed all but impossible to finalize the proposal, and for rereading and editing all the chapters. Your vision, resilience, and determination are incredible.

To Noel Weyrich, thank you for the endless edits and your wise guidance on how to shape the telling of my story. This book wouldn't be the same without your excellent contributions.

To Tim Burgard, Sicily Axton, and the rest of the HarperCollins Leadership editorial team, thank you all for your help and your commitment to excellence. And to Josh Gartrell, thank you for your friendship and for your brilliant work on the book's jacket cover. Sarge told me it's amazing!

Thank you to my entire extended family for your years of love and support: Erick, Dawn and Caleb Foster; Jeremy and Amber Wampler, Mark and Brandy Wampler, Paul and Shirley Wampler, Linda Wampler, Patti and Ron Evans, Sam and Sandra Foster, Jim Foster, Erbie and Mary Foster, Marcelo and Annie Seixas, Lelia and Luis Seixas, and Tiago and Zenith Seixas. To Erick, my brother, thanks for hanging with me from the start. To my cousins Mark and Jeremy, you've been like brothers to me since I was a toddler and always will be. Thank you both for the hand up when I needed it and the tough love when the times called for it.

Thank you to Elaine Dias for your calming help and guidance for our family.

To Mom and Dad, Peggy and Charles Foster, I cannot begin to thank you enough for all the sacrifices you've made on my behalf. The hours were long, and the road was bumpy, but we as a family never put much stock in style points.

To Juju and JP, Juliana and Jackson, thank you for being such remarkable children. I know it's not easy having a father who can't play ball with you or admire your artwork, as much as I wish I could. I thank God every day for bringing you both into our lives.

To Evie, my best friend, my wife, the love of my life, where do I begin? Without your tireless support, your fierce spirit and kind heart, none of this would have been possible. You are an incredible person, a wonderful mother, pure brilliance in motion, and I am so lucky to have you by my side.

ABOUT THE AUTHOR

CHAD E. FOSTER is a motivational keynote speaker, sales/finance leader, and inspirational change agent who works at Red Hat/IBM. He was the first blind executive to graduate from Harvard Business School's Program for Leadership Development and has engineered software thought impossible by technology giants. Despite going blind while attending college in his early twenties, Chad started at Accenture and has built a career in the technology industry where he has directed financial strategies and decisions resulting in more than $45 billion in contracts. He speaks to corporate audiences and professional athletes to help them develop resilience in the face of uncertainty. He lives with his wife and his two children in Atlanta.